23.95

COPY 1

D0913434

DYING FROM DIOXIN

DYING FROM DIOXIN

A Citizen's Guide to Reclaiming Our
Health and Rebuilding Democracy

By Lois Marie Gibbs and
the Citizens Clearinghouse for
Hazardous Waste

**BLACK
ROSE
BOOKS**

Montréal/New York
London

Black Rose Books No. AA246
Hardcover ISBN: 1-55164-085-6 (bound)
Paperback ISBN: 1-55164-084-8 (pbk.)
Library of Congress number: 96-79515

Canadian Cataloguing in Publication Data

Gibbs, Lois Marie
Dying from dioxin: a citizen's guide to
reclaiming our health and rebuilding democracy

ISBN 1-55164-085-6 (bound) -
ISBN 1-55164-084-8 (pbk.)

1. Dioxins—Health aspects. 2. Dioxins—
Environmental aspects. I. Title.

RA1242.D55G53   363.17'91   C96-900774-4

BLACK ROSE BOOKS

C.P. 1258	250 Sonwil Drive	99 Wallis Road
Succ. Place du Parc	Buffalo, New York	London, E9 5LN
Montréal,Québec	14225 USA	England
H2W 2R3 Canada		

To order books in North America: (phone) 1-800-565-9523 (fax) 1-800-221-9985
In Europe: (phone) 081-986-4854 (fax) 081-533-5821

Our Web Site address: http://www.web.net/blackrosebooks

A publication of the Institute of Policy Alternatives of Montréal (IPAM)
Printed in Canada

Contents

List of Tables

List of Figures

Acknowledgments

As grassroots organizers on a shoe-string budget, the people of Citizens Clearinghouse for Hazardous Waste are proud of the fact that we have celebrated more victories than we have accepted defeats. As an organization and as part of a grassroots movement, we've been able to succeed because of our working principle of mutual aid, and because of our outreach to those who have wanted to help but needed to be asked. This book is a perfect example of how CCHW and the movement it supports work. Many scientists, activists, leaders, and researchers contributed to this book by writing, reviewing, and helping to define its breadth. We want to thank and acknowledge each and every contribution. Without them, this book would not be as comprehensive, grounded, and useful as a tool for inspiring action to protect our health and our environment and to rebuild our democracy.

Special thanks go to Dr. Beverly Paigen, who not only wrote a large portion of the scientific chapters but also recruited other contributing scientists and provided direction for them. Dr. Paigen spent months reading thousands of pages of reports and translating very technical information into everyday language. Thanks to her work the public will be able to understand the severe threat that dioxin poses.

CCHW's organizing director Charlotte Brody, researcher and editor John Gayusky, and science director Stephen Lester also deserve special recognition and thanks. Each of them played a vital role in writing, researching, and factchecking. They also helped organize the roundtable meeting that brought leaders from diverse backgrounds together to help shape this book and CCHW's Stop Dioxin Exposure Campaign. Because we wanted to get this book out quickly, all three of them gave up months of their lives, sacrificing time with their families and friends, as well as a whole lot of sleep. It is an honor to be working alongside these three dedicated professionals.

Two organizations provided assistance above and beyond the call of duty: The Center For Biology of Natural Systems and Greenpeace.

Thanks to the scientists who contributed to this book, including Dr. Richard Clapp, Dr. David Ozonoff, and Tom Webster at the Boston University School of Public Health; Dr. Marvin Schneiderman, a retired scientist from the National Cancer Institute; and Dr. Frank Soloman at the Massachusetts Institute of Technology.

Thanks also to the researchers, organizers, and others who have directly contributed to this book, including Deborah Carpenter, CCHW; Lin Chary, Chicago, Illinois; Alicia Culver, Ralph Nader's Government Purchasing Project; Charlie Cray, Greenpeace; Elizabeth Crowe, Kentucky Environmental Foundation; Holger Eisl, Center for the Biology of Natural Systems; Richard Grossman, Provincetown, Massachusetts; Peter Kellman, labor organizer; Nina LaBoy, Bronx Clean Air Coalition; Marilyn Leistner, Times Beach, Missouri; Teresa Mills, Parkridge Area Residents Take Action; Tyion Miller, CCHW; Dr. Peter Montague, Environmental Research Foundation; Pete Sessa, Esq., Boston, MA; Barbara Sullivan, CCHW; Carol Van Strum, Tidewater, Oregon; Terri Swearingen, Tri-State Environmental Coalition; Joe Thornton, Greenpeace; and Craig Williams, Chemical Weapons Working Group.

Special thanks to the roundtable meeting participants, who sacrificed a weekend to come together and help define the Stop Dioxin Exposure Campaign, and later took the extra time to read the first draft of the book: Dollie B. Burwell, Warren County Concerned Citizens Against PCBs, North Carolina; Leslie Byster, Silicon Valley Toxics Coalition, California; Jackie Hunt Christensen, Institute for Agriculture and Trade Policy, Minnesota; George Coling, Rural Coalition, Virginia; Mark Cohen, Center for the Biology of Natural Systems, Queens College, New York; Dr. Paul Connett, Work on Waste, New York; Lorrie Coterill, Groups Allied To Stop Pollution, Texas; Charlie Cray, Greenpeace, Illinois; Alicia Culver, Government Purchasing Project, Washington,

D.C.; Joanne Eash, Concerned Citizens of Union County, New Jersey; William A. Fontenot, Louisiana Environmental Action Network, Louisiana; Sharon Golgan, People Against a Chemically Contaminated Environment, Arkansas; Diane Heminway, Citizens Environmental Coalition, New York; Madelyn Hoffman, Grass Roots Environmental Organization, New Jersey; Patricia Jackson, People for Community Recovery, Illinois; Hazel Johnson, People for Community Recovery, Illinois; Nina Laboy, South Bronx Clean Air Coalition, New York; Cynthia C. Laramore, Blacks on the Serious Side, Florida; Denise Lee, Blue Ridge Environmental Defense League, North Carolina; Esperanza G. Maya, People for Clean Air and Water, California; Teresa Mills, Parkridge Area Residents Take Action, Ohio; Calvin Mitchell, Center for Environmental and Community Justice, Indiana; Mike Morrissey, Save Our Cumberland Mountains, Tennessee; Karl Novak, Pennsylvania Environmental Network, Pennsylvania; Kim Phillips, Groups Allied to Stop Pollution, Texas; Lynn Pinder, Physicians for Social Responsibility, Washington, D.C.; John Pruden, Huron Environmental Activist League, Michigan; Virgil Reynolds, Save Our County, Ohio; Daniel Rosenberg, U.S. Public Interest Research Group, Washington, D.C.; Joe Schwartz, Physicians for Social Responsibility, Washington, D.C.; Alonzo Spencer, Save Our County, Ohio; Michael Sprinker, International Chemical Workers Union, Ohio; Lynn Thorp, Greenpeace, Washington, D.C.; Craig Williams, Chemical Weapons Working Group, Kentucky; Jane Melanie Williams, California Communities Against Toxics, California; Margaret Williams, Citizens Against Toxic Exposure, Florida; and Matthew Wilson, Massachusetts Campaign to Clean up Hazardous Waste, Massachusetts.

Finally, a second round of thanks to the roundtable participants and others who meticulously edited and added to the first draft of this book: Leslie Byster, Jackie Hunt Christensen, Mark Cohen, Alicia Culver, Rick Hind, Madelyn Hoffman, Mike Morrissey, Kim Phillips, Carol Van Strum, Tom Webster, and Matt Wilson. Thanks also to editor Sonia Shah and the rest of the South End Press collective.

It is a pleasure and an honor to work with the people named in these acknowledgements. I look forward to seeing all of them at the party to celebrate the victory of the Stop Dioxin Exposure Campaign.

—Lois Marie Gibbs and the
Citizens Clearinghouse for Hazardous Waste

Preface

By Lois Marie Gibbs
Executive Director and Founder, CCHW

In 1978, my neighbors and I discovered that our neighborhood in Love Canal, New York, had been built next to 21,800 tons of buried toxic chemicals. When we bought our homes, none of us knew that Hooker Chemical Corporation, a division of Occidental Petroleum, had dumped 200 tons of a toxic, dioxin-laden chemical called trichlorophenol and 21,600 tons of various other chemicals into Love Canal. We just knew we were getting sick. We knew there were too many miscarriages, too many birth defects, too many central nervous system problems, too many urinary tract disorders, and too much asthma and other respiratory problems among us.

But Hooker Chemical Corporation and Occidental Petroleum knew that the chemicals they had buried in the canal could damage the health of the people who lived in the surrounding neighborhood. In 1953, Hooker had filled in the canal, smoothed out the land, and sold it to the town school board for $1.00. The deed contained a stipulation that if anyone was harmed by the buried waste, Hooker and Occidental would not be responsible.

After we organized and won evacuation from Love Canal in 1980, I moved to Virginia to give my children a home safe from dioxin and other toxic contamination, and to start an organization that would help grassroots organizations fighting toxics in their own neighborhoods.

Until the United States Environmental Protection Agency (EPA) released the draft of its latest reassessment on the health effects of dioxin on September 13, 1994, I thought my move to Virginia had protected my children from further damage from dioxin. I didn't know we were feeding our children dioxin every time they drank milk or ate fish or meat. I didn't realize that the dioxin from a hazardous waste–burning aggregate kiln in Florida could end up in the milk of a cow or a new mother living in Michigan.

I was stunned when I read in the EPA report that "some of the effects of dioxin and related compounds, such as enzyme induction, changes in hormone levels, and indicators of altered cellular function, have been observed in laboratory animals and humans at or near levels to which people in the general population are exposed." The EPA report revealed that dioxin is much more prevalent—and, even for the population at large, much more dangerous—than previously reported.

You don't have to live next to Love Canal in New York or Mt. Dioxin in Pensacola, Florida, to suffer the effects of dioxin. The average boy, girl, woman, or man in the United States has enough or almost enough dioxin in his or her body to damage his or her health.

Because most of our dioxin exposure comes from the food we eat, we can't solve this problem just by eliminating the dioxin sources in our own backyards. The dioxin made by chlorine bleaching at Champion Paper in the backyards of families in Canton, North Carolina can end up in the ice cream eaten by residents of East Liverpool, Ohio. The dioxin emitted by the WTI hazardous waste incinerator in the backyards of East Liverpool can get into the hamburgers sold to North Carolina families.

The only way we can save our families from further exposure is to eliminate the sources of dioxin in everyone's backyard. Activists in East Liverpool, Ohio need to shut down that incinerator to protect not only themselves, but also the families in Canton, North Carolina. We need to save ourselves by saving each other.

That is what this book is for—to provide every reader with enough information to help stop dioxin exposure. Please don't get stuck in the science. Use it as a springboard for organizing and action. To save ourselves and to save each other, every community needs to have an active and committed group of residents working to shut down local sources of dioxin.

The first half of this book describes how dioxin is destroying the health of the American people. Most of what is written in the first half is a summary, in plain English, of the 2,400-page EPA draft report. We've also included important studies on dioxin published since the EPA report and a short history on the politics surrounding the science of dioxin.

The second half of the book describes how to reclaim our health by stopping our exposure to dioxin. You'll find advice on how to build an activist group, create a coalition, shut down an incinerator, convince your local government to buy only dioxin-free products, and more.

On April 28, 1995, forty people gathered in Arlington, Virginia, for the CCHW roundtable on dioxin. The purpose of this three-day event was to design the components of a national grassroots campaign to stop dioxin exposure. These components form the backbone of the second half of this book.*

The goal of our campaign is a sustainable society in which there is no dioxin in our food or breast milk because there is no dioxin formation, discharge, or exposure.

To reach this goal we aim to:

---

* Starting a campaign with a roundtable of citizen activists, scientists, and representatives of national organizations is a well-practiced CCHW tradition. The solid waste roundtable in 1986 resulted in the McToxics Campaign, a nationwide grassroots effort that reduced the use of styrofoam in the fast food industry.

1. Halt all incineration, including medical waste incinerators; municipal waste incinerators; hazardous waste incinerators; military waste incinerators; sewage sludge incinerators; and hazardous waste burning in cement and aggregate kilns, boilers, and industrial furnaces.

2. Expose and challenge all dioxin assaults on communities made up of low-income people and people of color.

3. Phase out industrial uses of chlorine, including its use in pulp and paper manufacturing and in PVC plastics, and include provisions for affected workers.

4. Identify more clearly the various sources of dioxin.

5. Determine the levels of dioxin in food and breast milk so that the progress of the campaign can be measured.

6. Promote safe, alternative jobs, products, and technologies.

Towards these ends, we will work to:

1. Build coalitions.

2. Educate the American people about dioxin and about how to stop dioxin exposure.

3. Keep involved communities connected.

4. Work together with labor.

5. Oppose all incineration, especially medical waste incineration.

6. Convince government, institutions, and consumers to buy only products that don't add to the dioxin levels in our food and our bodies.

7. Pass state pollution prevention laws.

In seeking to reduce dioxin exposure, we could blame the victim and get everybody to stop eating milk, fish, and meat and to stop breast-feeding their babies. Or we can use the same energy to stop the corporations who are filling our food and our bodies with dioxin. This book is our vote that we protect our health by reclaiming our right to safe food, safe water, and safe breast milk by shutting down the sources of dioxin.

In his book *Who Will Tell the People*, journalist and author William Greider writes:

> The American moment will begin ... when Americans find the courage to speak honestly again in the language of democracy.

We can't shut down the sources of dioxin without finding the courage to change the way government works. To begin this process of change, we have to create a national debate, community by community, on the nature of our government and our society. We have to explore how people became powerless as the corporations became powerful. We have to discuss why our government protects the right to pollute more than it protects our health. We have to figure out how to speak honestly and act collectively to rebuild our democracy.

For Environmental Justice,
Lois Marie Gibbs
Falls Church, Virginia
April 1995

About CCHW

The Citizens Clearinghouse For Hazardous Waste (CCHW) is a fourteen-year-old non-profit organization. Founded in 1981 by Lois Gibbs, leader of the campaign at Love Canal, CCHW is the only national environmental organization started and led by grassroots organizers. Thousands of people contacted Lois as her efforts and those of her neighbors captured national attention and proved for the first time that toxic hazards are not an abstract issue, but a reality that can exist in everyone's backyard. CCHW was created to respond with information, assistance, and solidarity.

In 1981, the main focus of CCHW was on helping community groups suffering from the effects of toxic dumps similar to Love Canal. Since then, CCHW has expanded it programs to match the expanding concerns of grassroots environmental groups. To date, CCHW has worked with over 8,000 community-based groups nationwide.

Science and Technical Assistance

Led since 1981 by Stephen Lester, a toxicologist with degrees from Harvard and New York University, CCHW translates scientific issues into plain language. Throughout CCHW's history we have helped thousands of activists understand technical and scientific information so they could articulate their concerns and have a voice in decisions that directly affect their lives. Our technical assistance program provides one-on-one work with community groups by reviewing and commenting on technical reports, cleanup plans, health studies, and alternative technologies. Through this work, communities come to understand, often for the first time, why their families are sick or what type of cleanup is needed to protect residents' health from further harm.

Organizing and Leadership Training

Led by Charlotte Brody, a registered nurse with thirty years of organizing and training experience in the civil rights, women's, and occupational health movements, CCHW provides organizing assistance to people by responding to telephone and written inquiries and by conducting trainings in communities. CCHW helps people start groups and keep them going. We have successfully connected individual groups with other community organizations to build strong, larger-than-local organizations.

CCHW has brought thousands of community activists together at regional training events called Leadership Development Conferences (LDCs) and Continuing Education for Organizers (CEOs). We train leaders on how to get their neighbors involved, how to develop winning strategies, and how to build democratic organizations.

CCHW brings grassroots activists, scientists, and representatives of labor unions and national environmental organizations together at roundtable meetings to discuss and engage in strategies to protect health and the environment. Roundtable participants are recruited from the leadership of grassroots organizations. They also include scientists concerned about the issue being discussed, labor unions, and national organizations working on the legislative aspects of the issue. It was from the roundtable process that the Stop Dioxin Campaign goals emerged.

Information Services

CCHW has published over sixty self-help guides to fill the needs of grassroots groups. In addition, we have developed an extensive library which includes government documents as well as files of information on corporations, and databases and a public computer link to other information that communities can use to win environmental justice. CCHW also publishes two periodicals: *Everyone's Backyard*, a quarterly magazine, and *Environmental Health Monthly*, a journal which reprints scientific articles on health and chemical exposures.

Board and Finances

CCHW is governed by a ten-member board. Five are grassroots leaders and five are professionals; four are men and six are women; four are people of color. Sixty-seven percent of our funding comes from foundations and the remainder comes from our members.

Highlights of CCHW's Accomplishments

Creation of Superfund

The federal Superfund Program, established in 1980 to clean up waste sites, was a direct outgrowth of CCHW Executive Director Lois Gibbs's work with her neighbors at Love Canal. Gibbs is commonly referred to by media and Congressional representatives as the "mother of Superfund."

Stopped New Hazardous Waste Landfills

CCHW's "from the bottom up" strategic grassroots effort has ensured that only one new commercial hazardous waste landfill has been built in this country since this effort began. The laws still permit such facilities, but people don't.

Kick Ash Campaign

CCHW assisted a grassroots effort to stop the deregulation of solid waste incinerator ash, which can be filled with dioxin and other toxics. The classification of ash as "special wastes" would have allowed ash disposal without testing for toxicity.

Solid Waste Project

Through community efforts to close cheap, unsafe solid waste disposal facilities and stop the construction of new facilities, grassroots leaders have increased the cost of solid waste disposal. These increased costs have persuaded many municipal and state governments to institute environmentally sound and less expensive methods of managing waste by mandating a percentage of waste to be recycled.

Superfund Technical Assistance Project

As a result of grassroots efforts, the federal government agreed to provide technical assistance grants to communities in amounts of up to $50,000. The grants are used to hire technical assistance providers to help communities evaluate cleanup options and site-specific environmental and health studies.

Right-to-Know Project

CCHW assisted in coordinating efforts to pass right-to-know laws, which give people access to information on chemicals being stored, disposed of, and released in their communities. CCHW worked with state grassroots groups to establish state government programs that provide local access to this kind of information.

Toxic Merry-Go-Round Campaign

With local grassroots organizations, CCHW helped develop strategies and regulations to halt the practice of taking waste that has been cleaned up at one site and dumping it on another community.

Race, Class, and Waste: The Connection

CCHW exposed the Cerrell report and other studies that document the deliberate siting of unsafe facilities in communities in which most residents are people of color, low-income, rural, Catholic, elderly, and/or without college educations.

McToxics Campaign

This campaign resulted from a national roundtable meeting about solid waste and excessive packaging. Roundtable participants decided that if we could convince McDonalds to stop using styrofoam, other fast foods companies would follow their lead. The three-year campaign was a success. McDonalds agreed to stop using styrofoam in packaging its products. In addition, during the campaign people nationwide organized efforts to stop the use of styrofoam in restaurants, schools, and public and private institutions.

Establishment of Local and State Protective Laws

CCHW has assisted groups in developing and then passing many laws to protect health and the environment. These include state Superfund programs; "bad boy" laws, which prohibit corporations with felony records from conducting business in a state; moratoriums on landfills and incinerators; and bans on the dumping of wastes produced in and transported from other states.

The History of Dioxin Exposure

1949 Accident at Monsanto plant in Nitro, West Virginia, exposes workers to dioxin.

1953 Accident at BASF plant in West Germany releases dioxin into two nearby communities.

1957 Dioxin isolated and identified as cause of workers' chloracne and illness. Outbreak of chicken edema disease kills millions of chickens in the southeastern United States. Chickens die after eating feed contaminated with dioxin-laced pentachlorophenol.

1962 to 1970 Agent Orange extensively sprayed in Southeast Asia during the Vietnam War.

1965 and 1966 Dow Chemical Company finances scientific experiments in which dioxin is applied to the skin of prisoners in Holmesburg Prison in Pennsylvania.

Mid-1960s Outbreak of reproductive and developmental effects noted in fish-eating birds of the Great Lakes.

1968 Approximately 1,800 people in Yusho, Japan, consume rice contaminated with PCBs and dibenzofurans. This and a similar accident in Yu-cheng, Taiwan (1979) provide graphic evidence of the effects of mixtures of PCBs and furans on humans.

1970 U.S. bans military use of Agent Orange in Vietnam.

1971 Dioxin-laced waste oil first used to control dust on roads in Times Beach, Missouri; town evacuated in 1983 after flooding spreads dioxin throughout entire community. Dioxin found to cause birth defects in mice.

1972 to 1976 Ah receptor theory of dioxin toxicity introduced and developed.

1973 Polybrominated biphenyls (PBBs) accidentally added to cattle feed in Michigan, killing hundreds of cows and forcing the slaughter of thousands of cattle headed for market.

1974 Dioxin detected in breast milk of mothers in South Vietnam.

1976 Hoffman-LaRoche trichlorophenol plant explodes in Seveso, Italy, exposing 37,000 people to a toxic cloud that contains dioxin.

1977 Dioxin found to cause cancer in rats. Federal court bans U.S. Forest Service use of dioxin-contaminated herbicides on national forests. U.S. commercial production of PCBs halted.

1978 Dioxin discovered at Love Canal in Niagara Falls, New York. Two hundred forty families evacuated in August; less than two years later another 740 families will be evacuated. Dioxin discovered in emissions of municipal waste incinerators.

Study by Kociba, et al., shows cancer in rats exposed to TCDD; will be used as basis for future EPA risk assessments for dioxin exposure levels.

Dow Chemical Company proposes "dioxin from fires" theory.

1979 EPA issues emergency suspension for some uses of 2,4,5-T after preliminary results of Alsea study show "overwhelming surge of miscarriages during the two months following the herbicide application."

EPA finds 40 parts per million dioxin in wastes at the Vertac plant in Jacksonville, Arkansas.

Yu-cheng rice oil contamination in Taiwan.

Association discovered between exposure to dioxin in phenoxyacetic acid herbicides and soft tissue sarcoma. Dioxin found to mediate hormones and their receptors.

1981 Capacitor fire in Binghampton, New York, contaminates state office building with PCBs and furans.

1983 to 1985 General public found to be contaminated with dioxin.

1985 EPA health assessment of dioxin released.

1986 Dioxin found in paper products due to chlorine bleaching used to make paper white.

1988 EPA begins first reassessment of dioxin.

1990 Banbury Center Conference on Dioxins sponsored by EPA and Chlorine Institute.

1991 First Citizens Conference on Dioxin held in Chapel Hill, North Carolina; organized to provide the public and grassroots activists with scientific information on the toxicity of dioxin.

Second EPA reassessment of dioxin begins.

NIOSH mortality study of U.S. chemical workers finds strong link between certain cancers and dioxin exposures.

1992 U.S.- Canadian International Joint Commission Sixth Biennial report calls for a phase-out of chlorine.

1993 Study by Bertozzi, et al., on residents of Seveso, Italy, discovers increased cancers in residents living near the plant at the time of the 1976 explosion.

1994 Second Citizens Conference on Dioxin held near Times Beach, Missouri.

1994 EPA releases draft reassessment document on dioxin; holds meetings in nine cities to obtain comments from the public.

CCHW kicks off Stop Dioxin Exposure Campaign.

1995 CCHW sponsors citizens roundtable meeting on dioxin. EPA Science Advisory Board meeting reviews draft reassessment document on dioxin.

—Adapted from Webster.

The Politics of Dioxin

Dioxin is the common name for a family of chemicals with similar properties and toxicity. Seventy-five different forms of dioxin exist, the most toxic being 2,3,7,8-tetrachlorodibenzo-p-dioxin, or TCDD. Dioxins are not deliberately manufactured. These chemicals are the unintended byproducts of industrial processes that involve chlorine, or processes that burn chlorine with organic matter.

The full story of dioxin and how it has come to affect the lives of virtually every man, woman, and child in the United States is a complex one. The toxicity of dioxin has been the subject of tremendous research and debate. At the same time, the political maneuvering by corporations who profit from its production and the failures of government to stand up to corporate pressure are an integral part of the dioxin story.

The dioxin story is the story of how science has failed to provide us with answers, how corporations control policymaking and decisions in our society, and how government is silenced.

The dioxin story includes coverups, lies, and deception; data manipulation by corporations and government as well as fraudulent claims and faked studies. For the public, it's a story of pain, suffering, anger, betrayal, and rage; of birth defects, cancer, and many other health problems. It's a story of money and power; of how corporations influence government actions and how this collusion affects the public.

But it's also an inspiring story of people saying, "No, I'm not going to take it any more"; of citizens getting involved and taking a stand against the corporations and the government that have injured them and lied to them; of people defending their families and communities; of enormous growth in grassroots resistance—new community groups, new leaders, and new heros and heroines who are taking back their health and their democracy.

The story of dioxin begins with chemical plant workers in Nitro, West Virginia; Midland, Michigan; Jacksonville, Arkansas; and western Germany, where dioxin-contaminated products were manufactured. It includes American soldiers and the people of Vietnam who were sprayed with Agent Orange, and people who were exposed during aerial spraying of forests in the Pacific Northwest. It also includes residents of Times Beach, Missouri; Jacksonville, Arkansas; Love Canal, New York; Pensacola, Florida; Seveso, Italy; and thousands of other communities around the world that have been contaminated with dioxin. It's a story that is continuing today as we struggle to stop the industrial processes that generate dioxins and release them into the environment.

Dioxin Research and the Chemical Companies

Although many companies have been involved with dioxin, three chemical companies have played particularly significant roles: Monsanto, BASF, and Dow Chemical. All three manufactured commercial products that were contaminated with dioxins. All three conducted health studies to evaluate the toxicity of dioxin. Corporations and government used the findings from these studies for many years to argue that exposure to dioxin had no long-term effects on human health.

In 1949, an explosion at the Monsanto chemical plant in Nitro, West Virginia, exposed many workers to the herbicide 2,4,5-T, which was contaminated with dioxin. Thirty years later, Monsanto scientists and an independent researcher who conducted many joint studies with Monsanto, Dr. Raymond Suskind, compared death rates among workers they said had been exposed to dioxin with death rates of workers who had not been exposed. When no differences between the two groups were found, Monsanto claimed that dioxin did not cause cancer and that there were no long-term effects from dioxin exposure (Zack, 1980). Additional studies released by Monsanto from 1980 to 1984 supported the general conclusion that there was no evidence of adverse health effects—other than chloracne, a severe skin disorder—in workers exposed in the 1949 accident.

Similarly, a chemical accident in 1953 at a BASF trichlorophenol plant in what was then West Germany released dioxin-contaminated chemicals, exposing workers and the nearby communities of Mannheim and Ludwigshafen. Again, scientists working for the company looked at cancer rates nearly thirty years later and found no differences between workers who were exposed during the accident and workers who were not exposed.

Both the BASF study and the first Monsanto study were released in 1980, shortly after researchers at Dow Chemical Company found that very low levels of dioxin caused cancer in rats. In response to the Dow study, BASF and Monsanto argued that their own studies conclusively showed that dioxin did not cause cancer in humans. Industry spokespersons used these results to challenge EPA efforts to regulate dioxin as a probable human carcinogen, arguing that humans respond differently than tested animals to dioxin exposure. People must be less sensitive, they argued. Otherwise, some evidence of cancer would have been found in these two "classic" studies. But when both studies were later re-examined, the methodology used in both was found to have serious scientific flaws.

Juggling Numbers, Part One: Monsanto

Evidence of inaccuracies in both the Monsanto and BASF studies was first revealed during the *Kemner v. Monsanto* trial, in which a group of citizens in Sturgeon, Missouri, sued Monsanto for alleged injuries suffered during a chemical spill caused by a train derailment in 1979 (Kemner, 1989).* While reviewing documents obtained from Monsanto, lawyers for the victims noticed that in one of the Monsanto studies, certain people were classified as exposed to dioxin, while in a later study, the same people were classified as not exposed (Hay, 1992).

The attorneys had compared individual information collected on cases in the first study—date of hire, year of birth, year of death, cause of death, and smoking history—to these same factors in the second study, and found that they didn't match.

Documents from the case reveal that Monsanto scientists omitted five deaths from the dioxin-exposed group, and also took four workers who had been exposed and placed them in the unexposed group. These changes resulted in a decrease in the observed death rate for the dioxin-exposed group, and an increase in the observed death rate for the unexposed group. Based on this misclassification of data, the researchers concluded that there was no relation between dioxin exposure and cancer in humans (Kemner, 1989).

The lawyers reworked the numbers and found that the death rate in the exposed group was 65 percent higher than expected, with death rates from certain diseases showing large increases. The death rate from lung cancer was 143 percent higher than

---

* The Kemner lawsuit against Monsanto had also revealed evidence that Monsanto willfully withheld from its customers and U.S. and Canadian agencies information about high levels of dioxin in Santophen, a common bactericide used in many household products, including Lysol brand disinfectant. The lawsuit concluded in 1988, when a district jury awarded $16.2 million in punitive damages to sixty-five plaintiffs. The jury ruled that Monsanto failed to warn residents of the presence of dioxin in the orthochlorophenol, a chemical used in wood preservatives, which was spilled in the train derailment (*Chemical Worker*, 1988). This decision was reversed on appeal on the grounds that a punitive award cannot be made in the absence of actual damages (Sanjour, 1994).

expected, the death rate from genitourinary cancer was 188 percent higher, bladder cancer deaths were up 809 percent, lymphatic cancer deaths were up 92 percent and the death rate from heart disease was 37 percent higher than expected (Kemner, 1989). Several other studies conducted by Suskind were published between 1980 and 1984. Suskind had proposed these studies to Monsanto in the wake of public attention given to the chemical release in Seveso, Italy, in 1976 (described in Chapter Seven). One of these, the Suskind-Hertzberg study, did not look at the original group of thirty-seven exposed workers, but instead looked at hundreds of other Monsanto workers from the Nitro, West Virginia, facility. Some of the same classification sleight-of-hand was performed in this study. Again, documents uncovered in the *Kemner v. Monsanto* legal case showed that there were twenty-eight cancer cases in the group of exposed workers and only two in the unexposed group. Suskind, however, found only fourteen cancers in the exposed group compared with six in the unexposed group (Suskind, 1984; Kemner, 1989; Van Strum, 1992).

Suskind also examined a group of thirty-seven exposed Monsanto workers during the four-year period following the 1949 accident. According to medical documents obtained by Greenpeace from the Kettering Institute in Cincinnati, Ohio, where Suskind worked, workers suffered "aches, pain, fatigue, nervousness, loss of libido, irritability and other symptoms... active skin lesions, [and] definite patterns of psychological disorders." He also found that all but one of the thirty-seven workers had developed chloracne, a severe skin condition. But in a report to Monsanto at the time, Suskind concluded, without further explanation, that "his findings were limited to the skin"; in other words, all other health effects of the dioxin exposure, other than chloracne, were not reported (Greenpeace, 1994). Out of these studies grew the industry claim that chloracne is the only long-term effect of dioxin exposure.

Juggling Numbers, Part Two: BASF

The study of BASF workers exposed to dioxin in 1953 was also found to have serious scientific flaws. BASF workers weren't convinced by BASF scientists' claim that there was no evidence of any health problems, other than chloracne, linked to dioxin exposure. They hired their own independent scientists to review the data. This independent review found that some workers who had developed chloracne, known to occur only in people exposed to high levels of dioxin, were included in the low-exposure or unexposed group in the BASF study. In addition, the exposed group in the BASF study had been "diluted" with twenty supervisory employees who appeared to be unexposed. When these twenty people were removed from the exposed group, significant increases in cancer were found among the workers exposed to dioxin (Yanchinski, 1989; Rohleder, 1989).

Corporate Studies and Public Health Implications

In February of 1990, when revelations of apparent fraud in the BASF and Monsanto studies came out in the *Kemner* trial, Dr. Cate Jenkins, project manager for the EPA Waste Characterization and Assessment Division of the Office of Solid Waste, alerted the EPA's Science Advisory Board (SAB) about these new findings. The SAB, which is an independent group of scientists from outside the agency, had recently completed a review of the cancer data on dioxin and concluded that there was "conflicting" evidence about whether dioxin caused cancer in people. They had recommended that the EPA continue to rely on the animal data (Jenkins, 1990).

This animal data, however, had been under attack since mid-1987, when, under pressure from industry, the EPA had acknowledged that it might have "overestimated" the risks of dioxin. The agency was prepared to lower its risk estimate for dioxin (*Inside EPA*, 1987). The key evidence leading to this conclusion was the Monsanto and BASF studies.

Jenkins asked the EPA to re-evaluate these proposed changes in light of the fraud allegations from the *Kemner* trial, and to conduct a scientific audit of Monsanto's dioxin studies. Instead, in August 1990, the EPA Office of Criminal Investigations (OCI) recommended a "full field criminal investigation be initiated by OCI." After two years of investigating the charges, however, OCI concluded that "the submission of allegedly fraudulent studies to the EPA were [sic] determined to be immaterial to the regulatory process. Further allegations made in the Kemner litigation appear to be beyond the statute of limitation" (Sanjour, 1994). In other words, OCI never completed its investigation because some of the alleged criminal activities had occurred more than five years earlier.

A documentary history of this investigation was prepared by William Sanjour, policy analyst for the EPA's Office of Solid Waste and Emergency Response and a long-time critic of the agency's hazardous waste policies. According to Sanjour:

> One gets the impression, on reviewing the record, that as soon as the criminal investigation began, a whole bunch of wet blankets were thrown over it. A finding of criminal fraud would have required first a finding that Monsanto's studies were scientifically flawed. Only an analysis by government scientists could have reached such a conclusion, and no EPA scientists were engaged in EPA's Monsanto investigation. None of the scientific groups in EPA, it seems, wanted to touch this hot potato, and no one in position of authority was instructing them to do so.

Rather than investigating all the allegations regarding Monsanto, the EPA actually spent two years investigating Cate Jenkins, the whistleblower whose memo, Sanjour says, precipitated EPA's crippled criminal probe of Monsanto (Sanjour, 1994).

The Monsanto and BASF studies were also the basis for arguments made by Syntex (USA), Inc., the company responsible for the cleanup of the dioxin-contaminated sites in Missouri. In 1986, Syntex prepared data that showed it could save 65 percent on cleanup costs if the soil cleanup standard was raised to allow

10 parts per billion (ppb) rather than 1 ppb. Syntex argued that the 1985 EPA risk assessment, which relied on animal studies to determine the 1 ppb cleanup standard, should be changed to take into account the Monsanto and BASF studies on workers, along with other evidence. Syntex also argued that by raising the standard, the agency (and other responsible companies) could save millions in cleanup costs (Commoner, 1994).

Subsequent studies on exposed workers at both the Monsanto plant in Nitro and the BASF plant in Germany have recently been published in scientific journals. (See Chapter Seven, "Dioxin and Cancer"). In 1991, the National Institute for Occupational Safety and Health (NIOSH) re-examined the causes of death in workers at the Nitro plant and found increases in all cancers among the exposed workers (Fingerhut, 1991). Similarly, in 1989, data on the BASF workers was re-examined and an increase in all cancers was found for workers with chloracne and with twenty or more years of exposure (Zober, 1990). The re-examination of these once "classic" studies provides strong evidence that the workers exposed to dioxin-contaminated chemicals in these two accidents did indeed suffer higher rates of cancer.

Playing with Fire: The Dow Chemical Company

Dow Chemical Company produced 2,4,5-T, an herbicide that was banned after being sprayed on forests in the Pacific Northwest, resulting in an alarming increase in miscarriages. Dow Chemical also produced Agent Orange, the herbicide that was sprayed on the jungles of Vietnam. Both herbicides are contaminated with dioxin during the manufacturing process.

In 1965, in order to evaluate the toxicity of dioxin, Dow conducted a series of experiments on inmates at Holmesburg prison in Pennsylvania. Under the direction of Dow researchers, pure dioxin was applied to the skin of prisoners. According to Dow, these men developed chloracne but no other health problems. However, no health records are available to confirm these findings, and no followup was done on the prisoners, even after several went to the EPA following their release. The former prisoners were seeking help from the EPA because they were sick, but the EPA did not respond (Casten, 1995).

In 1976, Dow began a series of experiments to evaluate whether animals exposed to dioxin would develop cancer. Dow chose very low exposure levels, perhaps hoping that the studies would show no toxic effects. Much to the company's surprise, the studies found increased rates of cancer at very low levels of exposure—the lowest as low as 210 parts per trillion (Kociba, 1978).

In 1978, the EPA began a study of miscarriages in Oregon. Evidence of increased miscarriages in areas of the Pacific Northwest that were sprayed with the herbicide 2,4,5-T was found (USEPA, 1979). Based on these findings, the EPA proposed a ban on this herbicide (Smith, 1979). But Dow brought its scientists to Washington, D.C., and created enough doubt and pressure that the EPA decided to only "suspend" most uses of 2,4,5-T. This enabled Dow to continue producing this poison. It was not until 1983 that all uses of this herbicide were banned.

In mid-1978, the Michigan Department of Natural Resources also found dioxin in fish in the Tittabawassee and Saginaw rivers. Dow discharged wastewater into these rivers and thus was an obvious suspect as the source of this contamination.

Dow responded in a most unusual way. The company proposed a new theory for the presence of dioxin throughout Midland, Michigan, the home of its largest plant. In November 1978, after an intense 4½-month effort that cost the company $1.8 million, Dow released a report called *Trace Chemistries of Fire* (Rawls, 1979). This report introduced the idea that dioxin is present everywhere, and has been around since prehistoric times. Its source, the report claimed, is combustion—any and all forms of burning—and no man-made chemicals need to be present to create dioxin in the environment (Dow, 1978). Dow concluded that the dioxin in the Tittabawassee and Saginaw rivers did not get there from Dow discharges, but rather from the "normal combustion processes that occur everywhere." Robert Bloom, director of research for Dow's production unit, was quoted at the time as saying, "We now think dioxins have been with us since the advent of fire. The only thing that's different is our ability to detect them in the environment" (Rawls, 1979).

Dow chose to release this report at a press conference rather than in the scientific literature, which is the usual route. The report was greeted with skepticism by many, as Dow clearly had a conflict of interest. But to others—including the Veterans Administration, which had just been sued by Vietnam vets for damages caused by exposure to dioxin in Agent Orange—this "news" was welcomed.

Today, this theory is still put forth, with a slightly different spin. Some industry representatives claim that forest fires are a significant source of dioxin exposure. According to industry, our real problem is not industrially produced dioxin but rather "natural" dioxin. But there is strong evidence refuting the combustion theory. The best evidence comes from time-dated measurements of dioxin levels in lake sediments (Czuczwa, 1984, 1985, 1986). Core sediments were taken from lake bottoms—both from lakes near industrial areas and from remote lakes—and measured for dioxin. According to the EPA, the results showed that dioxin was "virtually absent" until around 1940, when levels began to increase steadily before leveling off around 1960. This increase parallels the growth of chlorine-using industries.

Clearly, if dioxin had been created and distributed throughout the world since the "advent of fire," and if there were no significant contributions from man-made sources, then there would not be a sharp increase after 1940, when chemical companies such as Dow began to make products contaminated with dioxin.

In fact, using data from several studies, the EPA found that dioxin levels in sediment showed that man-made sources exceed natural sources by at least 100 to 1. This finding rules out any significant contribution from forest fires. Furthermore, fallout of airborne dioxin and dioxin precursors, such as pesticides and herbicides that are aerially sprayed and settle out on leaves and trees, can contribute to dioxin in emissions from forest fires.

Still other data refutes the "dioxin from fire" theory. Measurements of tissue from the bodies of nine Chilean Indians mummified for 2,800 years uncovered no dioxin (Ligon, 1989), and the bodies of two Eskimo women, frozen for 400 years, showed dioxins at only 15 percent of current levels, and no trace at all of very toxic 2,3,7,8-TCDD, the dioxin found in so many modern-day sources (Schecter, 1987). The data from the Chilean mummies is especially telling, since it represents cave dwellers who would have regularly burned wood indoors and thus were highly exposed to wood smoke. If dioxin was created by the "natural" combustion of wood, some appreciable amount of dioxin would have been found in the mummies.

According to the EPA, dioxin *can* be formed through natural combustion, but the contribution of natural combustion to dioxin levels in the environment "probably is insignificant" (USEPA, 1994d). While some minute levels of dioxin may have been evident in soil samples hundreds of years old, the critical point is that levels today are much higher, and that substantial levels did not begin showing up until after the birth of the petrochemical industry in the 1940s.

Dioxin in Paper

By the mid-1980s a new player had entered the dioxin arena: the pulp and paper industry. In tests conducted as part of the "National Dioxin Study," dioxin was found in wastewater discharges and paper products from pulp and paper mills that used chlorine as a bleaching agent.

In 1983, Congress had directed EPA to conduct this study in response to public concern generated by the discovery of dioxin contamination at Times Beach, Love Canal, Jacksonville, and other sites. The study was designed to investigate the extent of TCDD contamination nationwide. A major focus was on 2,4,5-trichlorophenol (2,4,5-TCP) production and manufacturing facilities, and associated waste disposal sites. The study included testing at pesticide formulation plants, pesticide use areas (rangelands, forests, rice fields, etc.), certain other organic chemical production facilities, and a limited number of combustion facilities. Several hundred soil and fish samples were also collected from sites around the country (USEPA, 1983).

Three years later, at the end of 1986, sensing that the study's results were, long overdue, Paul Merrell and Carol Van Strum, activists from the Pacific Northwest, filed a Freedom of Information Act (FOIA) request asking EPA for whatever results had been completed. When EPA failed to respond, they filed a lawsuit to get the information.

In a presentation at the First Citizens Conference on Dioxin, held in Chapel Hill, North Carolina, in 1991, Merrell described what happened next. "Greenpeace received a couple of documents leaked from EPA indicating that somehow, the pulp and paper industry had become involved in the National Dioxin Study and that there were signed secrecy agreements between EPA and the pulp and paper industry to keep the results of some sort of follow-up study secret. When we took those documents to court and raised a ruckus, suddenly EPA found thousands of pages of documents responsive to our FOIA request that they had previously denied even existed" (Merrell, 1992).

These documents included test results that clearly showed dioxin was present in wastewater and sludges coming out of pulp and paper mills nationwide. There was also evidence that dioxin was present in the pulp itself, which meant that dioxin would be in finished paper products as well. When the EPA released this data to Greenpeace, the agency tried to downplay the impact of the result by claiming that the data was "preliminary." But according to Merrell, the data had been validated and confirmed, and was sufficient to demonstrate that dioxin contamination was a serious problem for pulp mills.

The documents also provided strong evidence of collusion between the EPA and the paper industry to cover up the data (Weisskopf, 1987). This story is described in detail in a report written by Merrell and Van Strum and published by Greenpeace, called *No Margin of Safety*, which was based on some 65,000 pages of documents received from the EPA. The report, released in September 1987, turned pulp mill discharge of dioxin into a major public issue (USEPA, 1987; Van Strum, 1987; Weisskopf, 1987a).

As Merrell explained, "Shortly after *No Margin of Safety* was published, an official at the American Paper Institute, the lobbying arm of the American pulp and paper industry, read the report and got upset. This person, who [was] identified only as `Deep Pulp,' leaked about 300 pages of American Paper Institute documents to Greenpeace. These documents were the 'family jewels,' describing the paper industry's strategy for manipulating public information and opinion in an attempt to avoid liability and regulation.

"The leaked documents described the secret meetings with Lee Thomas, then the administrator of EPA, in which Thomas promised the industry that EPA would change its risk assessment for dioxin to get the pulp and paper industry off the hook for the dioxin levels in its wastewater and products." This proposed lowering of risk estimates prompted Cate Jenkins to call for a scientific investigation, as described earlier in this chapter.

According to Merrell, "The documents also described in great detail an agreement between EPA and the American Paper Institute, whereby EPA would send us... data but would accompany it with a letter saying that the data was preliminary and meaningless" (Merrell, 1992). The leaked documents also revealed that when the EPA informed industry that the paper mill test records were going to be released, the American Paper Institute hired a public relations firm, Burson-Marstellar, for a $100,000 retainer plus expenses. This firm was to publicize the fact that the EPA had written off the paper mill dioxin test results as "meaningless" in a letter to Greenpeace.

When Merrell and Van Strum received the leaked American Paper Institute documents, they filed them in their lawsuit, and asked the judge to allow them, through discovery procedures, to find out what was going on between the EPA and the paper industry. The judge, outraged at what was apparently happening behind the scenes at the EPA, issued the discovery order. He spoke bluntly about the agreement between the EPA and industry to "suppress, modify, or delay" the results of the National Dioxin Study, and encouraged the activists to find out more about what was going on.

After compiling depositions from EPA officials and the documents leaked by "Deep Pulp," Merrell summed up the situation by saying that the "EPA is not working for the public on the dioxin issue. They are working for the pulp and paper industry. If you approach this issue from any other standpoint, you'll run into stone walls all over the place" (Merrell, 1992).

Agent Orange and Vietnam Veterans

Many people first learned about dioxin by reading about the health problems suffered by soldiers who had been exposed to Agent Orange in Vietnam. This issue led to one of the most shameful episodes in dioxin's history. Between 1962 and 1971, the United States sprayed over 100 million pounds of Agent Orange over jungles in Southeast Asia (USEPA, 1982). This highly toxic herbicide was used to defoliate jungle areas that hid enemy troops during the Vietnam War. Soon after spraying, leaves would wilt and drop, leaving behind a barren landscape. Soldiers involved in Operation Ranch Hand, the most extensive of the defoliation missions, had the highest Agent Orange exposures.

By the late 1970s, veterans were reporting a variety of health problems—including cancer, numbness and tingling in the extremities, skin rashes, liver dysfunction, loss of sex drive, infertility, miscarriages, radical mood changes, weakness and birth defects in their children—which they felt were due to Agent Orange (Holden, 1979). By 1985, nearly 200,000 veterans had gone to the Veterans Administration (VA) to be examined for health problems they believed resulted from Agent Orange exposure (USGAO, 1986).

The American Legion and other veterans groups pressured Congress to investigate. Two sets of studies resulted. In August 1979, the Air Force "reluctantly" agreed to do a six-year study of 1,200 men who were involved in Operation Ranch Hand (Holden,

1979). Later that year, President Jimmy Carter signed a bill that ordered the VA to study the impact of Agent Orange spraying on Vietnam veterans. Despite its obvious lack of scientific credentials, the VA attempted to design this study itself, only to find that it could not. In 1983, the study was placed under the direction of the U.S. Centers for Disease Control (CDC).

At the CDC, the study was headed by Dr. Vernon Houk, director of the Center for Environmental Health and Injury Control. Three separate studies were designed by the CDC: (1) a study of the "general health effects" on military personnel of service in Vietnam (referred to as the "Vietnam Experience Study"); (2) a study of the long-term health effects of exposure to herbicides (referred to as the "Agent Orange study"); and (3) a study to determine veterans' risks of contracting selected cancers (Houk, 1989). The Vietnam Experience Study had three components: (1) an assessment of post service mortality; (2) a detailed health interview; and (3) a comprehensive medical, psychological and laboratory evaluation (CDC, 1987).

The study that received the most attention was the so-called Agent Orange study. This was the main study intended to determine whether veterans had suffered health problems due to exposure to Agent Orange. After spending three years trying to identify veterans who had been exposed to Agent Orange, the CDC announced, in June 1986, that the study couldn't be done because it was impossible to identify who had been sprayed and who had not. After briefings with White House staff, the study was canceled in 1987 without reaching any further conclusions (Booth, 1987).

The final piece of evidence examined by the CDC before it concluded that it could not go forward with the study was blood samples. In an effort to identify soldiers who were exposed to dioxin, the CDC compared dioxin levels in blood samples from soldiers who had been stationed in Vietnam with those in blood samples from soldiers who had been stationed elsewhere around the world. The CDC found the same levels of dioxin in both sets of samples (Booth, 1987; CDC, 1988).

Prior to cancelling the Agent Orange study, the CDC asked the National Academy of Sciences (NAS) to provide an independent assessment of whether the study could be completed. An NAS committee reported that the data available was more than sufficient to do a credible epidemiological study, but the CDC ignored this recommendation (Christian, 1992). The news that the CDC was abandoning its main study was greeted with great distress by veterans who had hoped the study would answer their many questions.

Later in 1987, a VA mortality study was released which found a 110 percent higher rate of non-Hodgkin's lymphoma and a 58 percent increase in lung cancer in Marines who served in heavily sprayed areas compared with those who served in areas that were not sprayed (Hilts, 1987). The high incidence of non-Hodgkin's lymphoma was confirmed in 1990 when the CDC released its "selected cancers" study (AL, 1990). Non-Hodgkin's lymphoma had been connected to dioxin exposure in other studies. However, the CDC dismissed this finding, arguing that the increase could not be linked to Agent Orange because of the lack of evidence of exposure. Veterans' groups and some scientists challenged this conclusion. The American Legion's John Sommer commented that the CDC had not actually measured exposure in any way. Instead, they had simply asked veterans whether they were exposed. This method is not generally considered reliable (*Science News*, 1990).

Then, in 1988, the CDC released the results of the study of the general health of Vietnam veterans (CDC 1988a, 1988b, 1988c). The CDC found that Vietnam vets suffered more depression, anxiety, and alcohol abuse than do other veterans, but they claimed to find no evidence of health problems related to exposure to Agent Orange. The CDC did report that Vietnam veterans had greater hearing loss, lower sperm counts, and more psychological problems, but claimed these differences were most likely due to "increased stress" (CDC, 1988b).

By this time, confusion about the studies' conflicting results—along with questions about how $63 million in government funds could be spent on studies that only concluded that no study could be done—prompted congressional hearings into these studies. In July, 1989, the Committee on Government Operations concluded that the CDC studies were "flawed and perhaps designed to fail" and that the government had "effectively used the CDC study to stifle any attempts to link Agent Orange to health effects" (Weiss, 1989). "Either it was a politically rigged operation, or it was a monumentally bungled operation," was the way Representative Ted Weiss (D-NY), chairman of the committee, described the canceled Agent Orange study (Yost, 1989).

During these hearings Dennis Smith, a CDC staff scientist, said that the CDC administrators had changed the design of the study so often and switched variables so frequently that the results were essentially meaningless. Smith said that researchers sometimes manufactured data to fill in gaps in the records, so that "at one point people lost track of what was true and what was false." When asked if he thought it was "impossible to link soldiers to exposure," as Houk had claimed, Smith said "that was completely false" (Yost, 1989).

The other set of government studies looking into Agent Orange-related health effects was conducted by the Air Force, which focused on the 1,200 soldiers exposed during Operation Ranch Hand. One of the first studies conducted by the Air Force was a mortality study, which was released in 1983. This study found that "fliers who were exposed to the toxic herbicides have not died at greater rates than other servicemen" (Hiatt, 1983). In 1984, the Air Force released "preliminary" results of a long-term study that found no relationship between herbicide exposure and adverse health effects, including cancer (Fox, 1984). The failure to find an increase in cancer was not surprising given the short time that had elapsed since exposure, the long delay in the development of cancer, and the young age of the veterans.

In 1988, the Air Force released an update on the health status of these soldiers. This time the Air Force found an increase in cancer of the skin, psychological damage, liver damage, cardiovascular deterioration, degeneration of the endocrine system, and birth defects in the children of the veterans. (USAF, 1988). The Air Force cautioned that Agent Orange could not be "confidently identified" as the cause of these health problems, but it also could not be ruled out. For the first time, a government agency had admitted that health problems suffered by Vietnam veterans could be linked to Agent Orange exposure.

Frustrated with the results of the CDC and Air Force efforts, the American Legion, in 1983, commissioned Columbia University to conduct a study of the health and well being of a random sample of Vietnam veterans. Drs. Jeanne and Steven Stellman led the study, which examined over 6,000 veterans. To help determine which veterans were sprayed with Agent Orange, the researchers filed a Freedom of Information Act request and obtained perimeter and helicopter spraying records for Operation Ranch Hand.

The results of this study were published in the December 1988 issue of *Environmental Research*, which was devoted entirely to the Stellmans' work. According to the American Legion, their results showed that "there is a significant correlation between Agent Orange, its basic ingredient dioxin, and medical disorders and birth defects" (Christian, 1994). These findings confirmed the results of the 1988 Air Force study.

Lost in the roller coaster ride of health studies and political maneuvering was the anguish and frustrations of the veterans and their families, who were suffering real health problems. In 1979 a series of class action law suits had been filed by veterans and their families against Dow Chemical and six other manufacturers of 2,4,5-T, one of the herbicides contained in Agent Orange.

The chemical companies submitted the BASF and Monsanto studies, the 1984 "preliminary" results of the Air Force Ranch Hand study, and other data to show that no adverse health effects other than chloracne were associated with human exposure to dioxin. U.S. District Court Judge Jack B. Weinstein relied on these studies in evaluating the veterans' claims.

The case never went to trial and was settled in 1984. Writing in *American Legion*, Ken Scharnberg details the outcome.

> According to the veterans' attorneys, Weinstein told them their evidence was weak and pressed them to accept an out-of-court settlement. Scant hours before

the trial was to begin, they caved in, settling for $240 million. Legal experts termed the amount as "nuisance value" money, cash the chemical companies gladly paid to prevent the case from being heard in court. Veterans and their families were to receive no more than $12,000 if the veteran was alive and no more than $3,500 if the veteran was dead (Scharnberg, 1992).

Many veterans were outraged at the size of the settlement. Thousands refused to accept the offer and opted out of the class action to sue individually. One veteran who refused to settle was Don Ivy, a Marine who later died from liver and pancreas cancer that he and his family believed was caused by Agent Orange. Shortly after his death in 1987, Ivy's wife filed a lawsuit against Diamond Shamrock, one of the manufacturers of Agent Orange. As part of this lawsuit, EPA chemist Dr. Cate Jenkins filed a 1991 affidavit which, according to the American Legion's Richard Christian, provided "an all-inclusive document linking the scientific and statistical facts compiled in dozens of recent studies proving beyond a shadow of doubt that a link exists between specific diseases and the dioxin contained in Agent Orange" (Jenkins, 1991; Scharnberg, 1992).

Adding further insult to the veterans, in 1989 the federal government dismissed the findings of the Columbia University study commissioned by the American Legion. In response, the American Legion, along with the Vietnam Veterans of America, filed suit against the federal government for failing to conduct a major ground troop study as mandated by Congress in 1979. This suit is still pending in the courts.

Then, in July 1993, the link between herbicide exposure and health problems was verified again in a study released by the National Academy of Sciences Institute of Medicine. The VA had requested that the NAS conduct an extensive study of herbicide exposure in Vietnam veterans. The NAS report concluded that there was a positive association between exposure to herbicides and soft tissue sarcoma, non-Hodgkin's lymphoma, Hodgkin's disease, chloracne, porphyria cutanea tarda (a liver disorder), as well as suggestive evidence for respiratory cancers (lung, larynx, trachea), prostrate cancer, and multiple myeloma (NAS, 1993). The NAS report led to the VA approving compensation for nine medical conditions resulting from exposure to herbicides used during the Vietnam War (Christian, 1994).

Retired Navy Admiral Elmo Zumwalt, who had been asked by the VA in 1989 to conduct a study of the Agent Orange issue, identified twenty-eight herbicide-related diseases that should be considered "service connected" and eligible for disability claims from the VA. Yet the VA still accepts claims for only nine diseases.

Today, there is little scientific doubt that exposure to Agent Orange is responsible for many health problems suffered by Vietnam veterans. There can also be little doubt that the U.S. government's response to the afflicted veterans has been at best callous, and at worst dishonest. The handling of the Agent Orange controversy by different parts of the government, including the Veterans Administration, the Centers for Disease Control, the Air Force, and the Army, has been remarkably similar to the deceitful manner in which other branches of the government have handled the question of whether dioxin causes health problems in workers and residents of communities contaminated by dioxin.

Admiral Zumwalt, who was responsible for ordering many aerial spraying operations in Vietnam and who lost a son to cancer which he believed was caused by exposure to Agent Orange, witnessed first hand the government's attitude towards the Agent Orange tragedy. In his 1989 study, Zumwalt found that "U.S. policy as decreed by the Office of Management and Budget was to avoid finding a correlation between Agent Orange and health effects. As a result, government studies had been manipulated to support that policy as had studies by dioxin-producing chemical companies" (Zumwalt, 1995).

Contributors

Stephen Lester, M.S.

Part One

The Science of Dioxin

Chapter One

The Toxicity of Dioxin

In this part of the book, we look at the science behind the health effects caused by exposure to dioxin. In doing this, we looked at many studies summarized by the EPA in their most recent reassessment of dioxin, released as a "public review draft" in September 1994. This draft consists of two documents; each is over 1,000 pages long and was published in three volumes. These two documents are generally referred to in this book as the "EPA Report." The first of these documents addresses the health effects of dioxin exposure (USEPA, 1994, 1994a, 1994b). The second focuses on the sources of dioxin and the levels to which the population has been exposed (USEPA, 1994c, 1994d, 1994e). The first document includes a chapter on risk characterization, which brings together the findings on both health effects and exposure levels, and describes the potential risks posed by dioxin. The entire draft is considered by the EPA to be a "scientific document that does not attempt to address policy or regulatory issues" (Farland, 1995).

Both the draft documents were reviewed in May 1995 by the Dioxin Reassessment Review Committee of the EPA's Science Advisory Board (SAB), a group of independent scientists from outside the agency. All of the exposure document and virtually all of the scientific chapters in the health document were endorsed by the SAB. The SAB's assessment was clear: The scientific basis of the draft report was sound and needed few, if any, changes. No new data was called for. The SAB disagreed with the EPA only in its interpretation of some of these scientific findings. So although the report was still in draft form, the scientific information used by the EPA was found to be accurate and scientifically sound. This is the data and information that we describe in this book.

We've taken what we believe is the most important information on the health effects of dioxin and, with the help of several scientists, "translated" this information into language that is accurate and easy to understand. We did not attempt to include everything in the EPA report, especially on the sources of dioxin and the levels of exposure. Instead, we focused on the most critical information about the toxicity of dioxin and how it affects the general public. At the end of each chapter we've included the names of the people who have contributed to its writing. References cited in the chapter are included at the end of the book.

Throughout this book, the term "dioxin" is used in two ways. When we are talking about laboratory studies with animals, dioxin means 2,3,7,8-tetrachlorodibenzo-p-dioxin, or TCDD, the most toxic member of the dioxin family. Virtually all animal studies have used this form of dioxin. When we are talking about exposure to workers, a community, wildlife, or domestic animals, we are referring to the whole family of dioxins and dioxin-like chemicals present in the environment. This includes all forms of chlorodibenzodioxin, including TCDD ("dioxins"), all forms of chlorodibenzofuran ("furans"), and all forms of chlorobiphenyls ("PCBs"). In the real world, TCDD is rarely found alone. Dioxin is almost always found as a mixture of its different forms.

The History of the EPA's Evaluation of Dioxin: The Scientific Issues

The scientific debate around dioxin has revolved around two primary issues: (1) Are people less sensitive to dioxin than animals? (2) Is there a "threshold" for exposure to dioxin? In other words, is there some level of exposure to dioxin at which no adverse health effects occur?

The current scientific evidence indicates that people experience many of the same toxic effects that animals experience. And the evidence is stronger than ever that dioxin causes cancer in people. The data suggests "a quantitative similarity between animal and human responses for some effects as well, including certain biochemical endpoints and carcinogenicity" (Webster, 1994).

The current scientific evidence also supports the conclusion that there is no threshold for the toxic effects of dioxin. More importantly, the debate around thresholds has been made moot by the finding that the average background "body burden" level, or amount of dioxin in the body, in the general U.S. population is close to levels that have been found to cause health problems. In other words, everyone has some amount of dioxin in his or her body, and the average level is already high enough to endanger health. So even if a threshold does exist, many people have already crossed it.

The First EPA Risk Assessment: 1985

The first evidence that dioxin causes cancer came from animal experiments completed in 1977. Cancer was found in rats at dioxin exposure levels as low as 5 parts per trillion (ppt) (Van Miller, 1977). Shortly afterwards, several other more extensive studies were conducted. The most important of these was a study published in 1978 by a team of scientists from Dow Chemical Company, led by Richard Kociba. Kociba found liver cancer in rats exposed to 210 ppt dioxin (Kociba, 1978). Finding cancer in animals at this low exposure level established dioxin as the most potent animal carcinogen ever tested. This result led to the general statement that dioxin is "the most toxic chemical known to man."

Kociba's finding generated scientific controversy and a call for more studies. The National Toxicology Program (NTP), a program of the National Institutes of Health which conducts animal studies to evaluate whether chemicals cause cancer, completed a major study of the carcinogenicity of dioxin in 1982. The

Table 1-1

"Acceptable" Daily Doses of Dioxin (pg/kg/day)

USEPA	0.006
State of California	0.007
Centers for Disease Control	0.03
US Food and Drug Administration	0.06
National Research Council of Canada	0.07
Germany	1-10
Netherlands	4
Canada and Ontario	10
World Health Organization	10
Washington State Department of Health	20-80

Source: Webster, 1994

NTP study also found cancer at levels similar to those in the Dow study in dioxin-exposed mice and rats (NTP, 1982a, b).

In 1985, the EPA published a scientific review of the health effects of dioxin, which served as the scientific basis for the dioxin risk assessments used for all EPA programs. The EPA review used the Kociba study as a basis for determining the risk of getting cancer from exposure to dioxin. The EPA's risk assessment led to the establishment of an "acceptable" daily dose of dioxin of 0.006 picograms per kilogram of body weight per day (pg/kg/day). For a 150 pound person, for example, this "acceptable" dose is equivalent to 0.42 picograms per day (USEPA, 1985). (See Appendix B for a conversion chart for picograms and other units used in this book).

Acceptable daily doses calculated by other countries and government agencies ranged from 0.007 to 20 pg/kg/day. (See Table 1-1). When the EPA's "acceptable" dose is compared to those of other countries, it's easy to understand why the EPA came under attack from industry. The EPA's risk estimate was the lowest of all.

The First Reassessment: 1988

Not surprisingly, a number of dioxin-generating industries and owners of dioxin-contaminated sites claimed that the EPA's "acceptable" daily dose estimate was too low, and that higher estimates were more appropriate and accurate. By 1987, the EPA's National Dioxin Study had found dioxin in the effluent of paper mills across the country and the paper industry had joined with other companies to pressure the EPA to reconsider its risk estimate. Shortly afterwards, the agency agreed to "reassess" the risks of dioxin and set up an internal "workgroup" of EPA staff to review the 1985 health assessment document and re-examine the data and reasoning.

Most of this reassessment centered around the different theories about how chemicals like dioxin might cause cancer. In the 1985 assessment document, the EPA chose to use the "linear multistage model," which is based on the assumption that cancer involves a sequence of irreversible stages and that the responses increase linearly with the exposure level. In other words, according to the EPA's model, there is no "threshold" of exposure to dioxin that is safe. Any exposure will increase the risk of getting cancer. The alternative theories all predicted an "acceptable" risk level for dioxin that was much higher than the value EPA calculated using the linear multistage model.

The EPA's workgroup could not agree on the best theory to use, so it decided to average the acceptable risk values predicted by the various models. This resulted in an "acceptable" risk value that was sixteen times higher than the EPA's original estimate. When the workgroup's unusual approach was reviewed by the EPA's Science Advisory Board, the workgroup was chastised. Several of the models used to calculate the average risk value directly contradicted one another. If one was right, then the other had to be wrong. The way to figure out which findings were correct was certainly not to average them. The Science Advisory Board found no scientific basis for revising the EPA's original acceptable risk value for dioxin. For the time being, the 1985 risk assessment had survived (Commoner, 1994).

A new challenge to the EPA's "acceptable" risk value came in 1989 from the pulp and paper industry, during a review of a water quality standard for the State of Maine. The paper industry decided to challenge the Kociba study that had linked dioxin exposure to cancer at very low levels. The original tissue samples used by Kociba were reviewed by a panel of "independent" scientists who examined each slide and voted on whether cancer was evident or not. The majority of these scientists concluded that there were 50 percent fewer tumors than in the original count. But the votes were far from unanimous, and some evidence of cancer was still present (Bailey, 1992). The EPA was not convinced by this divided review, and decided not to make any changes to its estimates. The 1985 risk assessment had survived another challenge.

In October 1990, the EPA and the Chlorine Institute, which represents companies that use or manufacture chlorine, cosponsored a scientific meeting entitled the "Biological Basis for Risk Assessment of Dioxins and Related Compounds," held at the Banbury Center on Long Island, New York. The purpose of the meeting was to review new data about how dioxin caused cancer in order to provide a "scientific" basis for EPA's risk assessment. This "new" data, which turned out to be over fifteen years old, showed that many of dioxin's effects involved binding to the "Ah receptor" of a cell. (See Chapter Four, "Dioxin Inside Our Bodies.") Most of the meeting focused on the role of the Ah receptor in mediating dioxin's toxic effects.

While most of the information introduced at this conference was not new, a highly controversial new theory was proposed. The theory suggested that a threshold for dioxin must exist since a certain number of Ah receptors must be occupied by dioxin for any biological effect to occur. In other words, people could be exposed to small amounts of dioxin without any observed effects. The conclusion that some participants drew from this theory was that the EPA's 1985 risk assessment was invalid and needed to be revised.

The Chlorine Institute hired a public relations firm, which issued a press packet falsely claiming that formal consensus on the threshold theory had been reached at the conference. This attempt to characterize scientific opinion and mislead the public was met with loud and angry protests from the scientific community, including researchers who had attended the meeting (Roberts, 1991).

The Second Reassessment: 1991

In spite of the scientists' protests, the damage had been done. Based on these "new" findings and the "consensus" that came out of the Banbury conference, the EPA's administrator, William Reilly, announced in April 1991 that the EPA would undertake a second reassessment of dioxin. Reilly, always willing to work with industry, was confident that this reassessment would reduce concerns about dioxin's toxicity. According to Reilly, the primary focus of the second reassessment was the idea of a threshold as discussed at the Banbury conference, and a new cancer model that took into account the presence of a threshold (Roberts, 1991a). News of the planned reassessment, including the notion that dioxin was much less toxic than previously thought, was widely reported in the lay press (Schneider, 1991; Gorman, 1991).

Meanwhile, the paper industry used the reported outcome of the Banbury conference to argue for relaxed dioxin water quality standards in a number of states. The Washington State Department of Health issued revised guidelines for fish consumption based on a tolerable intake in the range of 20–80 pg/kg/day, well above the EPA value of 0.006 pg/kg/day. No references to the scientific literature were given as a basis for this decision (Webster, 1994).

Yet barely five months into the EPA's second reassessment, new research by George Lucier and Chris Pottier of the National

Institute for Environmental Health Sciences suggested that there was no threshold for some of dioxin's effects (Roberts, 1991b). This and other research presented at the Eleventh International Symposium on Chlorinated Dioxins and Related Compounds, held in North Carolina in September 1991, seriously weakened the threshold theory. Other findings about dioxin emerged from this symposium, include the following:

1. Dioxin acts like a hormone, with effects that include neurotoxicity; immunotoxicity; and reproductive, developmental, and endocrine toxicity, including diabetes. Both people and animals showed similar sensitivity to these effects.

2. Additional evidence exists that exposure to dioxin at high levels for long periods of time causes cancer in people.

3. "Toxic equivalence" exists between various types of dioxin and dioxin-like chemicals. This means that the toxicity of other forms of dioxin (other than TCDD) and similarly structured chemicals such as PCBs and furans could be compared to the toxicity of TCDD. This was due to the similarity in how the toxic effects of dioxin and dioxin-like substances occurred in the body (HED, 1992). The concept of "toxic equivalence" was introduced and accepted by many conference participants (although not those representing dioxin-generating industries).

Much of the research that came out of the 1991 Symposium on Dioxin supported the findings of the EPA's original 1985 health assessment and provided the basis for the current reassessment, which was released by the EPA as a "public review draft" in September 1994 . In this report, the EPA pulled together the latest scientific evidence on dioxin and dioxin-like chemicals (PCBs and furans), including information on healths effects, sources, how people are exposed, and what happens to dioxin after it enters the body.

Several important findings came out of this reassessment report:

1. Dioxin-related health problems other than cancer may have more impact on public health than the cancer-causing effects of dioxin.

2. The EPA's conclusion that dioxin causes cancer in people was reaffirmed and strengthened by the data. The EPA estimated that at current exposure levels, the dioxin-related cancer risk is between 1 in 1,000 and 1 in 10,000. This risk level is 100 to 1000 times higher than the generally "acceptable" risk level of one in a million. This risk estimate makes dioxin the most important cancer causing chemical for the general population.

3. Dioxin accumulates in biological tissues, and the average level of dioxin in our bodies is at or just below levels that cause some adverse health effects.

4. The major route of human exposure is through ingestion of a wide variety of common foods containing small amounts of dioxin. This has resulted in the widespread exposure of the general public. Dioxin has been found in very high levels in human breast milk, the top of the human food chain.

5. Some health effects of dioxin (suppression of the immune system; reduced testosterone levels, which affects fertility; and reduced glucose tolerance, which increases the risk of diabetes) were found to occur "at or near levels to which people in the general population are exposed."

6. The principal sources of dioxin in the environment are combustion and incineration, chemical manufacturing, pulp and paper mills, metal refining and smelting, and "reservoir" sources (dioxin-contaminated soils and sediments) (USEPA, 1994).

One of the most striking findings of the EPA report is the significance of past dioxin exposures for public health. The report refers to levels of dioxin in the human body as the "body burden." According to the EPA, some adverse effects of dioxin exposure occur at levels slightly above average body burden levels currently found in the population, and "as body burdens increase within and above this range, the probability and severity as well as the spectrum of human non-cancer effects most likely will increase."

This means that, as a society, we have been accumulating dioxin and dioxin-like chemicals in our bodies. As a result, we are very close to "full"—our limit. It will only take small additional exposures to push some people's dioxin levels up to the level that will trigger adverse health effects. For most people, any additional exposure to dioxin, no matter how small, may lead to some adverse health effects. For others, cumulative exposures to dioxin may already have resulted in health problems. Problems such as infertility, diabetes, learning disabilities, childhood cancers, all of which are increasing in the United States, may be caused by the buildup of dioxin and dioxin-like substances in the body.

Dioxin's toxic effects reach beyond those who are directly exposed. Unexposed babies whose mothers and fathers were exposed earlier in life may suffer health damage as well. The EPA documents how dioxin causes damage that may not be "expressed" until much later in life. This damage can result in developmental and reproductive problems that may not be obvious until puberty or adulthood. Dr. Barry Commoner, Director of the Center for the Biology of Natural Systems at Queens College in New York, stated that we now know that the damage dioxin causes is "imprinted for life on the developing fetus" (Commoner, 1994a).

The EPA admits that it is impossible to state exactly how or at what levels of additional dioxin exposure people will respond, but the "margin of exposure (MOE) between background levels and levels where effects are detectable in humans... is considerably smaller than previously estimated." In other words, no amount of additional exposure to dioxin is safe.

The EPA report continues to undergo review. As mentioned earlier, the EPA convened a Science Advisory Board (SAB) meeting in May 1995 to review the draft reassessment report. All of the exposure document and virtually all of the scientific chapters of the health assessment document were approved by the SAB. The only disagreement was in how the EPA interpreted its scientific findings in chapter 9 of the health document. The SAB suggested clarification and amplification of some of the EPA's points, but no major overhaul was called for. One panel member characterized the SAB's concern about this chapter as simply needing "ripening" (Ozonoff, 1995).

The EPA anticipates releasing the final health and exposure documents in the fall of 1995. In hopes of delaying the release of the final report, scientists representing corporate interests have challenged the EPA's conclusions, but little of the scientific basis for these conclusions. Whether the EPA will give in to these challenges remains unclear at this time.

What we do know is that there's no need for us to delay taking action to stop or at least reduce some of the many ways we are exposed to this toxic chemical. We've waited long enough while the "experts" debated and asked for more data. There is no longer any doubt about the toxicity of dioxin and the dangers that it poses.

Dioxin is a dangerous chemical and a serious public health threat. No amount of additional exposure is safe. Every effort should be made to eliminate dioxin at its sources, rather than control it after it is produced. The EPA should stop trying to convince the public that an "acceptable" level of dioxin exists. We know that it doesn't. If the lessons of Times Beach, Love Canal, and Agent Orange teach us anything, it is that we should stop producing and using chemicals that will end up poisoning us.

Contributors

Stephen Lester, M.S.

Port Arthur, Texas. © *Sam Kittner*

Chapter Two

What Is Dioxin?

Dioxins are a family of chemicals with related properties and toxicity. There are 75 different forms of dioxin, the most toxic of which is 2,3,7,8-tetrachlorodibenzo-p-dioxin, or TCDD. When we use the term "dioxin" in this book, we usually mean all of the toxic forms of dioxin and dioxin-like substances.

Dioxin is not manufactured or used. Instead, it is formed unintentionally in two ways: (1) as a chemical contaminant of industrial processes involving chlorine or bromine, or (2) by burning organic matter in the presence of chlorine. Also, certain metal compounds can act as catalysts to increase dioxin formation in industrial processes using chlorine. Dioxin itself has no practical use.

Dioxin-like Substances: PCBs, Furans, and Brominated Substances

Some chemicals that are structurally similar to dioxin have "dioxin-like" behavior and toxicity. These chemicals include chloro dibenzo furans* also called polychlorinated dibenzofurans, PCDFs, or furans, chloro biphenyls* also called polychlorinated biphenyls, or PCBs, diphenyl ethers, and naphthalenes. There are

---

* We've written the names of these chemicals with spaces between the various parts to help the reader get familiar with the names, even though the correct spelling is to treat them as one word.

135 different furans and 209 different PCBs. Brominated substances—similar chemicals in which chlorine is replaced by bromine—may also have dioxin-like toxicity.

Not all dioxins, furans, and PCBs are equally toxic. Only 7 of the 75 dioxins are highly toxic, and only 10 of the 135 furans, and 11 of the 209 PCBs have dioxin-like toxicity. In this book, when we use the word "dioxin," we are usually referring to the activity of the 28 substances combined.

These 28 different dioxins, furans, and PCBs all exhibit similar toxic effects, caused by their similar activity inside our bodies: All bind to a portion of cell tissue known as the "Ah receptor" (See Chapter Four, "Dioxin Inside Our Bodies"). A key factor determining the toxicity of these chemicals is the size and shape of the chemical and how it "fits" with the Ah receptor to which it binds. Chemicals that fit tightly with the Ah receptor are more toxic than those that fit less tightly. TCDD has the tightest fit with the Ah receptor and is therefore the most potent and toxic dioxin. Chemicals that have about the same size and shape as TCDD fit almost as well with the Ah receptor and are almost as potent. Other chemicals that have a substantially different size and shape do not fit as well, or at all, and their toxicity is low or nonexistent. (See Appendix A, "The Chemistry of Dioxin and Dioxin-Like Substances.")

Toxic Equivalents

Dioxin-contaminated food, water, or soil can be contaminated with many different forms of dioxins, furans, and PCBs. Since some of these forms are more toxic than others, a way was needed to evaluate the toxicity of all dioxin-like substances present in a sample. Otherwise, tests on the sample would reveal only the risks posed by one form of dioxin, rather than the combined effect of all the dioxins and dioxin-like substances present. To do this, the EPA developed a two-step method for determining the total "toxic equivalent" present in a sample. In step one, the EPA uses a

Table 2-1

Toxic Equivalent Factors (TEF) for Dioxins and Furans

Chemical	TEF
2,3,7,8-tetra chloro dibenzo dioxin	1.0
2,3,7,8-penta chloro dibenzo dioxin	0.5
2,3,4,7,8-penta chloro dibenzo furan	0.5
2,3,7,8-tetra chloro dibenzo furan	0.1
2,3,7,8-hexa chloro dibenzo dioxin (3)	0.1
2,3,7,8-hexa chloro dibenzo furan (4)	0.1
1,2,3,7,8-penta chloro dibenzo furan	0.05
2,3,7,8-hepta chloro dibenzo dioxin	0.01
2,3,7,8-hepta chloro dibenzo furan (2)	0.01
octa chloro dibenzo dioxin	0.001
octa chloro dibenzo furan	0.001
all mono, di, and tri chloro dioxins and furans	0
other tetra, penta, hexa, hepta compounds without chlorines at the 2,3,7,8 positions	0

Note: Numbers in parentheses represent the number of different forms of dioxin or furan in this group.

Source: USEPA, 1994d

formula to convert the toxicity of all forms of dioxin into a common unit. In step two, the units are added up to give a total toxic equivalent.

(1) TEQ = [Concentration of dioxin] x [Toxicity Factor]

(2) Total TEQ = Sum of all TEQs present in sample

In the formula used for step one of this process, the most toxic form of dioxin, 2,3,7,8-TCDD is assigned a toxic equivalent factor (TEF) of 1. Each of the 17 toxic dioxins/furans is then assigned a "toxicity factor" that estimates its toxicity relative to TCDD. (Toxicity factors for PCBs have not yet been established.)

How Total Toxic Equivalent (TEQ) Is Determined

How are the toxic equivalents, or TEQs, that you find through-out this book calculated? When dioxin is measured in a sample (whether in food, tissue or incinerator ash, soil, air, water or any other medium), the amount of each individual form of dioxin present in the sample is measured. The concentration of each form is multiplied times the toxicity factor (TEF) for that form to yield a toxic equivalency to TCDD, the most toxic form of dioxin. The individual TEQs are then added together to give the total TEQ.

Let's take an example: How much TEQ dioxin is in a sample of ground beef? The following concentrations of dioxins/furans were found in a sample of ground beef. TEFs were obtained from Table 2-1 for each form of dioxin/furan. The TEQ for each form was determined by multiplying the concentration by the TEF. Total TEQ was determined by adding all the TEQs (column 3).

Form of dioxin/furan	Concen-tration (ppt)	TEF	TEQs (ppt)
2,3,7,8-tetraCDD	0.019	1	0.019
1,2,3,7,8-pentaCDD	0.062	0.5	0.031
1,2,3,6,7,8-hexaCDD	0.496	0.1	0.050
1,2,3,4,6,7,8-heptaCDD	1.157	0.01	0.012
2,3,4,7,8-pentaCDF	1.783	0.5	0.892
1,2,3,4,7,8-hexaCDF	4.846	0.1	0.485
TOTAL			1.489

From this calculation, the sample of ground beef has the total toxic equivalent of 1.5 parts per trillion of the most toxic dioxin, TCDD.

For example, a chemical that is half as toxic as TCDD is assigned a toxic equivalent factor of 0.5, and so on down to 0.001. The toxicity factor for each dioxin and furan is shown in Table 2-1.

To calculate the total toxic equivalent of a sample, the concen-tration of each dioxin/furan in the sample is multiplied by its assigned toxic equivalent factor. Then, all of the toxic equivalents are added up to give a total toxic equivalent concentration for dioxin as measured in TEQ.

But this is not the whole story. This is the calculation for dioxins and furans only, and these are the numbers you will see throughout the book. However, there are also eleven forms of PCBs that have dioxin-like toxicity. These PCBs are shown in Table 2-2. As Table 2-2 shows, the 3,3',4,4' positions on the PCB molecule are the key factors in determining toxicity for PCBs, just as the 2,3,7,8 positions are key for determining toxicity for dioxins and furans (see Appendix A for discussion of position of chlorines). The figures of dioxin and biphenyl in Appendix A also show that the 3,3',4,4' positions on the biphenyl molecule are similar to the 2,3,7,8 positions on the dioxin molecule. PCBs also have toxic equivalency factors, but scientists do not yet agree on exactly what these are. Therefore, the PCB contribution cannot be included in the total TEQ.

The EPA estimates that PCBs will double or triple the total TEQ. Dr. Arnold Schecter, editor of *Dioxins and Health*, agrees with this. He says that the TEQ for dioxins and furans should be multiplied by two to account for the contribution made by PCBs.

A s discussed above, the most important factor determining the toxicity of a dioxin or a dioxin-like chemical is its shape and size. The size and shape are further determined by the number and location of its chlorines. Most significant is whether there are chlorines in the 2,3,7 and 8 positions in the dioxin molecule, no matter how many other chlorines are present.

The significance of chlorines in the 2,3,7 and 8 positions can be seen from the toxicity factors in Table 2-1. Every dioxin or furan that has a toxicity factor has chlorines in the 2,3,7 and 8 positions. This includes octa (8) chloro dioxins that have chlorines at every position, including the 2,3,7 and 8 positions. Note that dioxins and furans usually have similar toxicities when their chlorine atoms are similarly placed.

Table 2-2
Dioxin-Like PCBs
3,3',4,4'- tetra PCB
3,4,4',5 - tetra PCB
2,3,3',4,4'- penta PCB
2,3,4,4',5 - penta PCB
2,3',4,4',5 - penta PCB
3,3',4,4',5 - penta PCB
2,3,3',4,4',5 - hexa PCB
2,3,3',4,4',5'- hexa PCB
2,3',4,4',5,5'- hexa PCB
3,3',4,4',5,5'- hexa PCB
2,3,3',4,4',5,5'-hepta PCB
Source: USEPA, 1994d

Physical Properties

Dioxins, furans, and PCBs share many physical properties, several of which determine how they will behave in the environment. Dioxin and dioxin-like chemicals are not very water soluble. They have a low vapor pressure, which means that they do not turn into gas very easily. They have a high octanol water partition coefficient, which means that in the body they do not get excreted in urine, but instead remain in the body, dissolved in fat. This property of dioxin's chemicals means that they will accumulate in the food chain.

Dioxins also have a high organic carbon coefficient, which means they stick to the organic components of soil and water and tend to stay where they end up in soil or mud, such as at the bottom of a lake or river. A small portion of dioxin transfers from water to air by turning into a gas. And finally these chemicals can be destroyed by sunlight but only when several other conditions are just right.

In sum, dioxins are mostly bound to particulate matter (dust) in the air and water. They bind to organic matter when they are in soil, sediment, water, and air. They are poorly soluble in water and very soluble in fat, meaning that they will bioconcentrate (build up in the body). Once bound to particulates, they do not leach (dissolve and wash away) or volatilize (turn into gas) easily. Dioxins with four or more chlorines in their molecular structures are extremely stable, which means they do not break down or change into other substances. The only significant breakdown is by photodegradation (breaking down by sunlight), but this occurs very slowly. Once in the air, they can be removed by photodegradation, by being washed out by rain, or by settling out as dust. In soil, these chemicals either get buried in place or erosion carries them into bodies of water. Sediment on lake, river, and ocean bottoms is often the final environmental reservoir of dioxin and dioxin-like substances.

Environmental Fate

Waste burning is the major source of dioxin and dioxin-like chemicals in the air. Dioxin can exist in the air as vapor (gas) or can be bound to particles, depending on the vapor pressure. Usually they are bound to particles, and because the particles are small, they can stay in the air for a long time and can be carried long distances. (See "Dioxin Fallout.") This is the major reason why dioxin is found everywhere in the world.

Dioxin in the air as a gas can photodegrade, but most airborne dioxin is bound to particles such as ash from incinerators. Bound dioxin does not photodegrade very easily either because the particle shields the dioxin from the sun or because the particle contains chemicals that inhibit photodegradation. Eventually, airborne dioxin

Times Beach, Missouri

Times Beach was a small suburban community situated on the Meramec River, 17 miles west of St. Louis, Missouri. Times Beach had been a flood plain used for farming, and by the early 1970s, "The Beach," as the natives called it, had a population of 1,240 people and two growing mobile home parks. It also had over 16 miles of very dusty roads.

To control the dust, the city contracted with waste oil hauler Russell Bliss to spray the roads at will during the summers of 1972 and 1973. This was thought to be a bargain at only 6 cents per gallon of oil.

Ten years later, on November 10, 1982, a reporter for a local newspaper informed our city clerk that Times Beach was possibly among the sites sprayed with waste oil containing dioxin. This news was followed by a call from the U.S. EPA confirming the reporter's information. We were told it would be as long as 9 months before any soil testing could be done.

Chaos broke loose. The residents immediately recalled that the roads had turned purple after being sprayed. The spraying had resulted in an awful odor, birds had died, and newborn animals succumbed shortly after their birth.

No one in our immediate area was familiar with dioxin or the possible effects of this chemical. When information on dioxin did come in, none of it was comforting. The EPA finally announced testing should be done immediately because of the number of people exposed.

The residents had taken up a collection and contracted with a local laboratory to do private testing, since we did not want to wait months for the EPA to do its testing. It seemed to us that the EPA accelerated their testing program only after they learned we were conducting our own tests.

On December 5th, a day after the first round of sampling was completed, the community suffered the worst flood in its history. Many barely escaped with their lives.

On December 11th, residents still burdened with the worries about their flood-damaged homes learned that another of their fears had been confirmed. The results of the private testing had been made public. Dioxin was definitely present, but the levels and extent of the contamination were not yet available.

Twelve days later, residents received what we now call our Christmas message. "If you are

is either washed out by rain (wet deposition) or settles onto the ground (dry deposition).

Dioxins that are not washed away will remain in soil for a very long time. They do not move down into groundwater, because they are not water soluble. It is estimated that it takes between 25 and 100 years for half of a given concentration of dioxin in the soil to degrade. Dioxin in the very top three millimeters of soil is slowly degraded by sunlight, but the soil just below the surface is not affected. For example, the dioxin concentration in the top one-eighth inch of the contaminated soil at Times Beach, Missouri, was decreased 50 percent by photodegradation over a sixteen-month period, but the dioxin concentrations below the surface did not decrease at all. Dioxin below the top inch did not migrate upward to vaporize into the air, and it did not migrate downward with the rain—it just stayed put (Freeman, 1987).

Since dioxin tends to persist in the soil, landfills or any other incidences of burial in soil create major environmental reservoirs for dioxin. However, dioxin can be mobilized by oil or chemical leachates that contain organic solvents such as benzene or toluene because dioxin can dissolve in these substances, (which are almost always found in landfills). In Love Canal, New York, dioxin moved long distances, from the canal into storm sewers and a nearby creek, because it was dissolved and carried by oil-like chemical leachate.

The sediment at the bottom of lakes, rivers, and oceans represents another major environmental reservoir of dioxin. Dioxin gets into bodies of water by settling out from the air, by being washed out of the air by rain that flows into the water, by erosion and by being directly released into the water by industries such as pulp and paper mills. Since dioxins do not easily volatalize from water into air, and photodegration is slow in water unless organic solvents are present, dioxin tends to remain in water just as it does in soil. Once in the water, dioxin sticks to particles and organic matter and eventually settles to

in town it is advisable for you to leave and if you are out of town do not go back." Those brave souls who continued with clean-up from the flood found some of their family members breaking out in rashes; others became ill. Some of us were sure the rashes were caused by dioxin, but others felt they came from a combination of mud and cleaning fluids.

It was August, 1983 before the first offers were made to begin the buy-out of residents. The offers were low, and threats of condemnation within 30 days if offers were not accepted caused resentment on our part. Homeowners spray-painted their offers on their homes as TV cameras focused on them. Offers eventually improved.

I cringe when someone says "Dioxin never hurt anybody." Dioxin has harmed everybody who has come in contact with it. For us it has meant red-lining by insurance companies, marital discord, a type of forced bankruptcy, loss of property values, community, neighbors, friends, identity and security, and, most of all, loss of our health.

—*Marilyn Leistner*
(The Orange Resource Book, 1995)

Dioxin Fallout

On May 18, 1995, the Center for the Biology of Natural Systems (CBNS) released a report showing that dioxin creates a toxic chemical fallout problem. Dioxin emitted from 1,329 U.S. and Canadian sources—mostly incinerators that burn municipal, medical, and hazardous waste—spreads through the air over the entire continent. Dioxin-contaminated dust travels very long distances, often covering more than 1,000 miles before settling, the study found.

According to Dr. Barry Commoner and Dr. Mark Cohen, the primary authors of the report, these findings explain why most human exposure to dioxin is due to contaminated dairy foods and beef, even though few farms and ranches where these foods are produced are located near dioxin-emitting sources.

Specifically, the CBNS report showed how dioxin reaches the Great Lakes. According to the report, only half of the dioxin deposits in the Great Lakes come from any of the dioxin sources within 300 miles of the lakes. The rest comes from as far away as Texas and Florida. The bottom. A disturbance of these sediments, however, can cause dioxin to re-enter the environment. Dioxin can be released, for example, through the dredging of a harbor.

From the sediments where they settle, dioxins are ingested by organisms and enter the aquatic food chain. As large fish eat smaller fish who have eaten even smaller fish and so on, dioxin concentrations in the aquatic food chain become larger and larger. The total dioxin in the large fish is the sum of the dioxin in each of the smaller fish that the large fish ate.

The environmental fate of PCBs is similar to that of dioxin. PCBs are also stable and persist in the environment. They are often bound to particles. There are, however, some important differences. PCBs are somewhat more reactive than dioxin, so they are more easily broken down by sunlight, air, and bacteria. PCBs in the air are in the vapor phase 90 percent of the time.

They are washed by rainfall out of air and may slowly evaporate back into air.

Brominated dioxin-like compounds behave similarly to their chlorinated dioxin cousins. They are relatively immobile in the environment, are bound to particulates and organic matter, and are significantly removed only by photodegradation. They are resistant to leaching in soil and to volatilization.

The EPA has found that ingestion of dioxin in food is the primary means by which people are exposed to dioxin. Dioxin in the air settles onto soil, water, and plant surfaces. When dioxin-laden plants or feed are eaten by sheep, cattle, or chickens, the dioxin ends up in the animal's fatty tissues. The same thing happens to fish living in dioxin-contaminated bodies of water. When people eat meat from these animals, dioxin can move up the food chain and enter the human body. Dioxin is not taken up through the root systems of plants and vegetables. It can, however, be moved by erosion or by surface water runoff into a body of water, where it enters the aquatic food chain.

report dispelled the myth that dioxin emissions are only a problem in the immediate vicinity of their source.

CBNS used computer models developed by the National Oceanic and Atmospheric Administration (NOAA) and data on weather patterns obtained from NOAA to estimate dioxin dispersion and transport from the sources they identified. The model was modified to take into account the different forms of dioxin and was verified by comparing results from air samples collected in Dorset, Ontario, with those predicted from the model. The actual results and the results predicted by the model were only slightly different.

This means that the model can be used to predict dioxin levels from a single or multiple sources at different locations. The CBNS database includes 954 individual sources by name, address, and estimated annual dioxin emissions. Another 375 sources are grouped by state or province (Cohen, 1995).

In summary, dioxin and dioxin-like substances are very persistent in the environment. They do not break down and are not metabolized by bacteria. Once they enter the environment, the primary way they are removed is by a very slow process of photodegradation or by disappearing into environmental reservoirs, such as lake or river sediments or the soil of landfills. These environmental reservoirs are not necessarily permanent. And dioxin enters the human body through the food chain, where—just as it does in the environment—it tends to stay put.

Contributors

Dr. Beverly Paigen
Stephen Lester, M.S.
Marilyn Leistner

Chapter Three

Where Dioxin Comes From

Dioxin and dioxin-like chemicals are found everywhere in the world in water, air, and soil. This distribution suggests that the sources of these chemicals are multiple and that they can travel long distances. Dioxin gets into the environment from industrial air emissions, wastewater discharges, and disposal activities, and from the burning of material containing chlorine. Once in the air, dioxin is mostly washed out or settles onto soil, plants, and water. It moves up the food chain, and some of it ends up in people's bodies.

Most of the information on sources of dioxin used in this book comes from USEPA estimates described in the exposure document portion of the agency's second reassessment of dioxin (USEPA, 1994d). There are many limitations—fully acknowledged by the EPA—with the data that is discussed later in this chapter. A major problem is the failure to include estimates of emissions from any industrial/chemical manufacturing plants, despite knowledge that these sources generate dioxin. According to the EPA, no emissions data exists for these types of facilities. In fact, the EPA's estimates are one of only two comprehensive efforts to identify and estimate dioxin emissions from different sources. The EPA's estimates are useful as a starting point for discussion, but our use of them is not meant to imply endorsement.

The second comprehensive effort to identify and estimate emissions from dioxin sources (which also lacks estimates of emissions from industrial/chemical manufacturing plants due to the same absence of available data) was conducted by the Center for the Biology of Natural Systems at Queens College in New York. The center, headed by Dr. Barry Commoner, has been on the leading edge of research on incineration and dioxin issues since 1981. The CBNS study, released in May 1995, looks only at air and water emissions. CBNS's total national estimate of dioxin air emissions was similar to the EPA's estimate. The EPA estimated that there were 9,300 grams toxic equivalent (TEQ) dioxin released per year into the air; CBNS's estimate was 7,800 grams TEQ/year. (See Chapter Two, "What Is Dioxin?" for an explanation of toxic equivalents.) Some individual source estimates were different, due to the availability of more up-to-date data and use of different research methods on the part of CBNS (Cohen, 1995). In this book, we rely mostly on the 1994 EPA estimates, although where the EPA made no estimate for a particular source, we have used the CBNS estimate.

An overview of how much dioxin is released in the U.S. each year from sources where the EPA made estimates is shown in Table 3-1. Each of these quantities is given as a toxic equivalent equal to the most toxic dioxin—2,3,7,8-TCDD. This table does not include dioxin already present in the environment.

The numbers given in Table 3-1 are estimates—or, to be more accurate, guesses—of the amount of dioxin released into the environment from sources on which the EPA had data. Some are very good guesses, based on many measurements, and some are fairly wild guesses. To give some appreciation for the quality of the guesses, the EPA has provided a high and low estimate for each source. For estimates that are very good (i.e., estimates in which the EPA has "high" confidence), the range between the high and low estimates is a factor of two. Examples are the numbers for dioxin in water and in commercial products. If the guess is based on reasonably good information with some rather large data gaps, confidence is "medium", and the range is about five-fold, as in the number for

Table 3-1
U.S. Dioxin Emission Estimates for Sources Included by EPA in Grams of Toxic Equivalents/Year

Emissions into media	Release (TEQ)	% of total	Range (low)	Range (high)
Air	9,300	80	3,300	26,000
Water	110	1	74	150
Land/landfill	2,100	18	1,000	4,500
Commercial products	150	1	110	220
TOTAL	11,660	100		

Note: The average value of TEQ released is determined by using the geometric mean, or a weighted average.

Source: USEPA, 1994d

dioxin emission falling onto soil. If the guess is based on only a small number of measurements, the confidence is "low", and the range is much greater, with about a ten-fold difference between high and low estimates, as in the number for air.

As the table shows, the most dioxin, by far, is released into the air. However, dioxin releases into water, although they account for only 1 percent of total dioxin releases, are very significant to human health, because dioxin concentrates so readily in the aquatic food chain, ending up in the fish we eat. Disposal of dioxin into the land, mostly in landfills and as wastewater sewage sludge that is spread on farmland, can also pose significant health risks because of the movement of dioxin out of these disposal sites. Most of the dioxin included in the landfill estimate is dioxin-contaminated ash generated by incinerators.

Four major industrial practices produce and release most of the dioxin that is present in the environment, and account for our subsequent exposure to dioxin and dioxin-like chemicals. The biggest by far is the practice of burning chlorine in the presence of certain precursors to dioxin (substances that can lead to the formation of dioxin), such as organic matter or lignin in wood or chemicals with similar structures. In incinerators, the amount of dioxin formed by burning seems to largely depend on the chlorine content in the waste that's burned. Chlorine in incinerators comes from such sources as polyvinyl chloride (PVC), vinylidene chloride (plastic wrap), chlorinated solvents, paint strippers, and pesticides. Chloride ions found in salt and food can also contribute to the formation of dioxin in incinerators, but these sources do not form dioxin as easily, or in as large quantities as do industrial chlorine sources.

The second major source of dioxin is the paper and pulp industry, which uses chlorine to bleach paper. Dioxin forms when chlorine reacts with organic lignin in wood. Lignin is the "glue" that holds trees together. In tests done as part of the National Dioxin Survey, dioxin has been found in pulp and paper mill sludge, in wastewater released from these plants, and in paper products.

The third major source of dioxin is chemical manufacturing of commercial products that contain chlorine, such as PVC, chlorinated solvents, and pesticides. The fourth major source is industrial plants, such as metal smelters and refineries and cement kilns. All of these sources have a common denominator—they all use chlorine in some capacity.

How Dioxin Is Formed

Dioxin is formed during industrial processes involving chlorine or when chlorine and organic (carbon-containing) matter are burned together. Three conditions are required for dioxin to form: (1) a source of chlorine (or bromine); (2) the presence of dioxin precursors such as organic matter or aromatic hydrocarbons ("ring" compounds like benzene); and 3) a thermally or reactive environment where these materials can combine. In general, the more chlorine present, the more dioxin is formed. This suggests a good control strategy: reduce the amount of chlorine and the amount of dioxin generated will be reduced.

Chlorine Chemistry

Dioxin is formed during the manufacture of hundreds of chlorinated chemicals called organochlorines. These products, such as the chlorophenol herbicides 2,4-D and 2,4,5-T, are made by mixing chemicals in carefully defined reactions. These reactions involve "ring" compounds such as chlorinated benzenes, chlorinated phenols, and other compounds. (For an example of ring compounds, See Appendix A, "The Chemistry of Dioxin and Dioxin-like Substances.") Virtually without exception, when these reactions occur, dioxin and dioxin-like substances are unavoidably generated.

According to the EPA, dioxin has been found not only in the anticipated benzene, phenol, and other ring compounds, but also in non-ring substances such as the solvents trichloroethylene

Carver Terrace, Texarkana, Texas

Carver Terrace is a subdivision located one and a half miles west of downtown Texarkana, Texas, a city of 32,000 people. Residents of Carver Terrace are African American, with an average income between $10,000 to $20,000 per year.

In 1903, the National Lumber and Creosote Company began operating a wood treatment facility on sixty-two acres on what was then the edge of town. In 1938, the land and the facility were sold to the Wood Preserving Corporation, which was eventually acquired by the Koppers Company. Koppers continued wood preserving operations until 1961, when the facility was closed and the land sold. In 1964, the seventy-nine single-family homes of the Carver Terrace development were built. Jimmy "Sonny" Fields was an original resident of the community. He said, "There was always a smell. It gets strong at times. But nobody knew it was a hazard to their health. People who were getting sick back in the earlier days never knew what was wrong, or the cause, until these test results began to leak out. The first test was taken before 1980, but was never made public."

The State of Texas and Koppers found that soil and groundwater beneath Carver Terrace were contaminated with common wood preserving chemicals, such as pentachlorophenol (PCP), arsenic, and creosote. Thirty-five on-site contaminants have been identified, including 770 parts per trillion of dioxin in the soil. Carver Terrace resident Talmadge Cheatham recalled an EPA official dressed in a moonsuit coming to his home to investigate the black goo that had backed up into his bathtub. Cheatham said, "Well, you certainly have to protect yourself, but what is supposed to protect me and my family?"

Five million dollars in federal funds were allocated to buy out and relocate the residents of Carver Terrace in the summer of 1990. But the evacuation came too late for community leader Patsy Oliver and her family. Patsy died in 1993. Her mother had died in 1989, and her daughter had died just a few weeks before Patsy.

Patsy's wish was "that the Environmental Protection Agency would do what it was mandated to do—protect all people and make polluters pay for the harm they do."

—*Sonny Fields*
(Newman, 1994)

(TCE) and tetrachlorethylene (perc), chlorinated dyes, and pesticides. Dioxin seems to be generated in all aspects of chlorine chemistry from the generation of chlorine gas to all "downstream" processes whenever chlorine is present (Thornton, 1994). Some organochlorine compounds are used as chemical "intermediates"—chemicals that help produce a product, and do not end up in the final chemical product. Still, dioxin has been found in some of these final products, such as the pesticide parathion or nitrophenol. Dioxin will thus end up either in the product chemical, in process residue, or pollutants released into the air or waste water. The extent of dioxin formation in the thousands of chlorinated products made is unknown. What is clear is that dioxin is formed from a large number of complex chemical reactions involving chlorine.

How Dioxin Gets into Air

Perhaps the most important source of dioxin is the process of burning. Dioxin forms when material or waste containing chlorine and dioxin precursors are burned. Table 3-2 shows approximate air emissions of dioxin from different sources in the United States. The major contributors are medical waste incinerators—primarily small incinerators at hospitals and laboratories—and large municipal waste incinerators. For some sources, the EPA made no estimate, as indicated in the table. These sources are placed in the table in their estimated order of importance. The ranges for each source are shown in Figure 3-1.

Dioxin Formation in Incinerators

Incinerators of all kinds provide an ideal environment for dioxin production. Most of the research on dioxin formation in incinerators has been done with municipal solid waste or household garbage incinerators, which provide plenty of chlorine sources and precursors to dioxin, and sufficient heat to drive the reactions that produce dioxin. As is evident in Table 3-2, all types of incinerators produce some amounts of dioxin. Dioxin is formed not just in old or poorly operated incinerators, but even in the most sophisticated state-of-the-art incinerators.

According to the EPA report, the pattern of the forms of dioxin found in the environment—with more forms having a higher number of chlorines—is the same pattern produced by incineration.

Studies at several garbage incinerators have showed that dioxin is not formed in the furnace itself, but rather sometime afterwards, while the emission gasses cool as they leave the incinerator through the emissions stack. The temperature at which the most dioxin forms is between 280 and 400 degrees Celsius. This means that contrary to some claims, a hotter temperature in the furnace is not effective in controlling or eliminating dioxin formation. Dioxin is still formed in incinerators with very high furnace temperatures in the 1200 degree Celsius range.

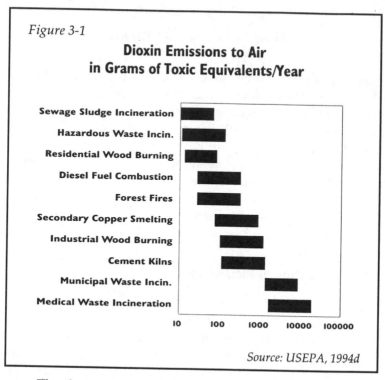

Figure 3-1

**Dioxin Emissions to Air
in Grams of Toxic Equivalents/Year**

Source: USEPA, 1994d

The electrostatic precipitator, an air pollution control device which removes dust particles from stack gasses, operates at the temperature range perfect for dioxin formation. Ironically, solid waste incinerators with electrostatic precipitators for air pollution control are the greatest individual source of dioxin emissions.

The chemistry of how dioxin is formed during burning is not clear. There are three main theories.

1. Dioxin is present in the waste from the beginning. However, comparisons of measurements of dioxin in the waste that's burned and in stack emissions show that the dioxin in the waste stream is not enough to account for the amount coming out of the stack.

2. Dioxin is formed from natural sources whenever any organic matter and chlorine are present in the waste stream, and are then burned together. Laboratory experiments have shown that formation from natural sources is possible under certain conditions, but the yield is extremely low.

Table 3-2

Air Emissions of Dioxin
in Grams of Toxic Equivalents/Year

	Emissions (TEQ)	Percentage
Medical waste incineration	5100	53
Municipal waste incineration	3000	31
Cement kilns	350	4
Wood burning	360	4
Coal burning	0-300*	3
Secondary copper smelting and refining	230	2
Iron Sintering	230**	
Chemical manufacturing	Not estimated	
Home and building fires	Not estimated	
Forest fires	86	1
Transportation Vehicles		
Diesel fuel	85	1
Leaded gasoline	1.9	<1
Unleaded gasoline	1.3	<1
Hazardous waste incineration	35	<1
Sewage sludge incineration	23	<1
Petroleum refining	Not estimated	
TOTAL	9800***	100

* The EPA said that the TEQ from coal burning was no more than 300 grams/year, based on the failure to detect any dioxin in the emissions at one plant. CBNS provide an estimate of 200 grams/year.

** The EPA did not estimate the emissions from iron sintering plants, so this estimate is from CBNS (Cohen, 1995).

*** The total emissions differ from the EPA's estimate of 9300 because the EPA did not include estimates for coal or iron sintering plants.

Source: USEPA, 1994d

3. Dioxin is formed from dioxin precursors in the waste stream that are broken down and rearranged as dioxin during the burning process. This appears to be the source of most dioxin.

The issue of how and why dioxin is formed through burning is important because industry scientists have used different theories—such as Dow Chemical's "trace chemistry of fire" theory, which attributed dioxin formation to "natural" combustion—to minimize and dismiss concerns about the toxicity of dioxin. (See the Introduction, "The Politics of Dioxin.")

How Dioxin Gets into Water

The largest source of the dioxin found in water is the wastewater discharged from the pulp and paper industry's bleaching process. The amount of dioxin from this source is known with a reasonable degree of certainty, although in recent years, data provided by the industry indicates that the level of dioxin in discharge of pulp and paper mills has decreased (NCASI, 1993). Sewage treatment plants, chemical manufacturers, and metal refineries also discharge dioxin into the water, but the quantity from these sources is unknown because of the lack of testing. Dioxin also gets into water from the air, when airborne dioxin is washed out by rain into water or onto soil, which can erode into streams and lakes. (See Table 3-3.)

Reservoir Sources of Dioxin

Dioxin and dioxin-like chemicals are very stable and thus stay in the environment for long periods of time. They concentrate in the food chain and a certain percentage ends up in us.

However, most of the dioxin released into the environment ends up in what are called "reservoirs." These reservoirs are a serious concern because dioxin can be mobilized from them and

Table 3-3
Wastewater Discharges of Dioxin in Grams of Toxic Equivalents/Year

Sources	Discharges (TEQ)
Paper and pulp	110
Sewage treatment plants	Not estimated
Chemical manufacturing	Not estimated
Metal refineries	Not estimated
TOTAL	110

Source: USEPA, 1994d

reenter the general environment. The EPA estimates that reservoir sources contain between fifteen and thirty-six times the amount of dioxin typically generated in one year. This means there is a great potential for future exposure even if we stopped all dioxin pollution today.

Reservoirs of dioxin include the aquatic sediments at the bottom of streams and lakes, landfills and contaminated industrial and waste sites, and fallout from air onto soil and vegetation. How likely is it that the dioxin in these reservoirs will re-enter the environment? The dioxin at the bottom of oceans, lakes, and rivers does get covered up by the deposition of new sediments each year. However, if the bottom is stirred up by the dredging of a harbor or river, or by a storm or a flood, a substantial re-entry of dioxin into the environment will occur.

Dioxin reservoirs in the hundreds of dioxin-contaminated industrial and hazardous waste sites across the country pose a substantial threat to people. Sites such as the Escambia Wood Treatment site in Pensacola, Florida; the Vertac trichlorophenol plant in Jacksonville, Arkansas; and the Diamond Shamrock plant in Newark, New Jersey; have all been identified as sources of dioxin contamination that has spread off-site into surrounding neighborhoods. Dioxin spreads from sites like these via groundwater, surface water, dust dispersion, and evaporation, and also by direct contact as people walked on these sites before they had been identified as a threat. Dioxin contamination from these sites can also be spread by cleanup activities and by trucks and cars that travel to and from the contaminated areas. Many of these sites are on the EPA's Superfund list of the worst hazardous waste sites in the country. And in some instances, the EPA or the responsible company has "cleaned up" the site by burning the wastes, which creates yet another source of dioxin. Dioxin in landfills can also re-enter the environment when solvents such as oil, degreasing agents, or organic solvents dissolve dioxin and carry it out of a landfill (HWN, 1992).

Another reservoir of dioxin is created by the fallout of airborne dioxin onto soil and vegetation. This fallout gets covered each year by the following year's deposition (clean fallout as well as more dioxin fallout), but construction, serious erosion, and other earth-moving activities can cause dioxin near the ground's surface to re-enter the environment. Dioxin deposited on forest leaves by airborne fallout re-enters the environment during forest fires and brush fires.

How Good Are the Estimates?

Estimating the amount of dioxin released into the environment involves many uncertainties. To get an idea of how accurate the estimates are, scientists have attempted to independently measure the amount of dioxin that has been deposited from the air onto the ground over a wide area, and then to compare this amount to known dioxin emissions. If we truly know all the sources of dioxin, and if the estimates of the emissions of dioxin from these sources are reasonably accurate, these two values—total deposition and total emissions—should balance. This comparison between dioxin deposits and dioxin emissions is called a "mass balance" evaluation.

Mass balance evaluations for dioxin have been completed in Sweden, Great Britain, and the United States. In Sweden, the deposition of dioxin in rural areas was measured and found to be 5 nanograms TEQ per square meter per year. This amount was multiplied by the total land area of Sweden and then compared to the amount of known emissions of dioxin. The rate of deposition was 10 to 20 times higher than could be accounted for by all known dioxin emissions in Sweden. In other words, the known dioxin sources could account for only 5 to 10 percent of the deposition rate. This was quite a serious discrepancy.

What could be the explanation? The major possibilities are that (1) there are some large unidentified sources of dioxin; (2) the emissions from known sources are much higher than estimated; (3) dioxin in reservoirs is re-entering the environment; (4) most of the dioxin is coming from outside Sweden; or (5) some combination of the above is occurring.

A similar study was carried out in Great Britain. The deposition rate was found to be 13 nanograms TEQ per square meter per year, quite a bit higher than in Sweden. This study also showed that the deposition rate was 10 times higher than could be accounted for by all the sources in the country. Both Sweden and Great Britain are small countries, so it is easy to think that the excess dioxin might be coming from outside the country. But what about the United States?

The EPA did a dioxin mass balance evaluation for the U.S. but had data on deposition from only one study, which measured the deposition rate at a rural site in Indiana. The deposition rate there was found to be between 1 and 2 nanograms TEQ per square meter per year. For urban areas, EPA estimated dioxin deposition to be between 2 and 6 nanograms TEQ per square meter per year. This estimate was based on five European studies that measured urban deposition rates: 13 nanograms (Great Britain), 5 nanograms (Sweden), 4–30 nanograms (Germany), 8 nanograms (Germany), and 7–36 nanograms (Germany). The EPA calculated a total U.S. deposition rate of between 20,000 and 50,000 grams TEQ per year.

When this estimate is compared with EPA's estimated air emissions, found in Table 3-2, of 9,800 grams per year, it's clear that, as with the estimates made in Sweden and Great Britain, the rate of deposition does not balance with known emissions. The U.S. deposition total, although an estimate, is much higher than known emissions. The EPA's estimate of total air emissions is only half of the estimate of the lowest deposition rate (9,800 grams TEQ per year compared with 20,000). Although there may be many reasons for this discrepancy, one simple conclusion that can be drawn is that we can account for only a fraction of total dioxin emissions.

Depositions vs. Emissions: The Missing 50 Percent

D ata from the mass balance evaluations suggests that some sources are putting out more dioxin than current estimates lead us to believe, that there are some unidentified sources, or that dioxin is re-entering the environment from its reservoirs. These reasons for comparatively low emissions estimates are consistent with several facts. The EPA made no estimate of dioxin emissions from chemical manufacturing plants and has not included them in its estimates of total emissions, as shown in Table 3-2. Yet in its report, the EPA has clearly identified dioxin in nearly 100 pesticides, in many chlorinated solvents and intermediates, and in dyes and pigments. According to the EPA, there was no available data on the dioxin emissions from these sources. The EPA also acknowledged that it had no information on dioxin emissions from plastics production facilities.

These chemical manufacturing facilities need to be included in the overall assessment of dioxin sources. It is unclear why the EPA chose to make no estimate of their contribution to dioxin emissions. Even if the EPA had no readily available data, it could easily have gone out and done the necessary testing, or required companies to do the testing themselves. Instead, the EPA has asked companies to "voluntarily call in" with emissions data so that the contributions from these sources can be estimated (USEPA, 1994f). This effort has little chance to succeed. To get a complete picture of how much dioxin is released from these types of facilities, the EPA must require industrial processing plants and other known sources of dioxin to measure how much dioxin they are releasing and make this data available to the public. The EPA has also failed to estimate contributions from reservoir sources. Hundreds of dioxin-contaminated industrial and hazardous waste sites exist around the country. Most of these sites have not been cleaned up, and they remain a major threat to their surrounding communities. Another example of the contribution of reservoirs to total dioxin emissions can be found in the Great Lakes. Recent research has shown that the Great Lakes, which has been the endpoint, or "sink," for PCB dumping, has now become a source of dioxin. Dioxin-like PCBs in the Great Lakes are re-entering the environment due to natural biological and physical activity as well as man-made disruption (Smith, 1995).

Another source that the EPA has overlooked is the contribution of dioxin from accidental fires in homes and buildings in which wiring, flooring, siding, molded furniture, and other materials containing chlorinated plastics provide the ingredients to generate dioxin. There are almost 700,000 structural fires each year, according to 1992 U.S. Census statistics (USDOC, 1993). Several studies document that burning PVC plastic results in dioxin formation (Thiessen, 1989; Christmann, 1989). Some researchers estimates that this source may contribute as much dioxin to the environment as medical or municipal waste incinerators (Thornton, 1995). Military waste incinerators, industrial furnaces and boilers, high-rise apartment incinerators, and backyard open-pit burners are still other sources not considered in the EPA report.

The emissions estimates are not the only side of the EPA equation subject to criticism. The EPA's estimate of dioxin deposition is based on only a single study conducted in Indiana. Additional studies are needed to verify this estimated deposition rate. The EPA estimated an urban deposition rate based on European studies, but chose one of the lowest possible deposition rates from this group of studies. The EPA's deposition rates may also be off because of losses that are not easily measured, such as dioxin emissions that are blown out to the ocean or broken down by sunlight. It is also difficult to evaluate how much dioxin is re-released through evaporation from contaminated surface or water or dust from contaminated surface soil.

Without additional, more complete, and more accurate studies, it is impossible to be certain which of these explanations accounts for the large discrepancy between the estimated emissions of dioxin, and the estimated deposition of dioxin in the environment. What is certain is that levels of dioxin present in the environment are even higher—and more dangerous—than the emissions reports suggest.

Contributors

Dr. Beverly Paigen
Stephen Lester, M.S.

Chapter Four

Dioxin Inside Our Bodies

How does dioxin get into our bodies? Once inside, where does it go? How long does it stay? And how does it affect our bodies and make us ill?

Many of the answers to these questions come not from experience with humans, but from experiments with laboratory animals such as mice and rats. Because these experiments are designed to answer specific questions, scientists have to make some aspects of the experiment artificial. Scientists do this by controlling the environment where the animal lives. Animal studies are designed so that scientists know exactly how much of a chemical the animals are exposed to, by controlling their air, water, and food. And animals are exposed to only one chemical at a time. People do not live in such controlled environments.

How much dioxin scientists give to lab animals varies greatly from one experiment to the next, as does how they give it: in their water or air in a pre-measured oral dose, or mixed in food, or applied to the skin. People mostly take in dioxin in their food and water, and in some circumstances through their skin or just by breathing.

Love Canal, Niagara Falls, New York

From the early 1940s until 1952, an abandoned canal in Niagara Falls, New York, was used by Hooker Chemical (owned by Occidental Petroleum) as a dumping ground for 21,800 tons of hazardous waste. Included in these wastes were 13 million pounds of lindane, 4 million pounds of chlorobenzenes, and about a half a million pounds of trichlorophenol used in the manufacturing of herbicides such as 2,4,5-T.

In 1954, an elementary school was built on the perimeter of the canal; the playground was built directly above the dump. Over 400 children attended the school. Houses and apartments built around the dump became homes for about 900 families. Barbara Quimby grew up at Love Canal. She remembers playing on the dump that they called "quicksand lagoon." She and her friends would poke sticks into and skip rocks across the black sludge that surfaced. Mothers had to call Hooker to ask how they should treat the chemical burns their children received from playing at the site. One child was burned over 70 per-

Retaining Dioxin, and Getting Rid of It

The actual amounts of dioxin used in laboratory experiments and found in the environment are very small. That doesn't mean that the amounts are insignificant, since small amounts of a powerful agent can have large effects. Scientists have worked out many ways to detect even a low level of dioxin once it enters the body, using very sensitive tests to follow its path. No matter how it enters the body, dioxin travels throughout the body in the bloodstream. But dioxin stays in the blood for only a short time. That's because dioxin does not dissolve very well in water, the major component of blood and of almost all the tissues in the body.

Dioxin does dissolve very well in organic or "oily" media. The fat in the body is such an "oily" substance, so dioxin preferentially ends up in fatty tissue. It also accumulates in the liver, because there are proteins in the liver that bind tightly to dioxin. Dioxin remains in the tissue for many years. According to the EPA, the half-life of dioxin (the amount of time it takes for half of a given amount of dioxin to break down) in people ranges from seven to eleven years (Pirkle, 1989).

The body gets rid of dioxin by metabolizing, or breaking it down, in the liver into compounds that are much more water soluble, which are then excreted. The process of breaking down chemicals like dioxin also renders the chemical less harmful. Since these changes happen very slowly in both people and laboratory animals, however, dioxin stays in the body for a long time. This is another reason why even small amounts of dioxin can be significant: Low doses build up in the body over time.

The rate at which dioxin is eliminated from the body differs from individual to individual. One key factor is the amount of body fat a person has. Because dioxin is stored in the fat, people with more body fat will store more dioxin. Furthermore, dioxin stored in fat will be eliminated more slowly from the body. Another key factor in determining the elimina-

cent of his body when he placed chemically reactive "hot" rocks in his pocket. They caught fire when they rubbed against each other as he played.

In 1977, an investigation of the site found that "fume odors from the site are evident at all times." The study also found that, over the years, drums of chemicals nearest the surface of the canal had become increasingly corroded and exposed, resulting in a massive leakage of contaminated liquid from the canal into the neighboring area. One of the chemicals leaking from Love Canal was dioxin.

Residents were told in 1978 that because dioxin adheres to soil it would not travel through groundwater and therefore there was no cause for alarm. This assurance was quickly discredited when cleanup workers discovered that, when mixed with chemical solvents, dioxin does move freely through soil with groundwater. Additional testing was then done throughout the community. Dioxin was detected blocks away from the dump in soil more than three feet below the surface.

It was then discovered that the cleanup activities spread dioxin-tainted dust throughout the neighborhood. Vegetation, from hedges and flowers to twenty-five-year-old oak trees, died throughout the neighborhood. As if trying to prove that the air was dangerous, a bird fell right out of the sky and died at the foot of a state environmental investigator who was walking the site "listening" to residents' concerns about the dead vegetation.

The health problems suffered by Love Canal residents were horrifying. In a nine-block section of the community, nine out of sixteen babies born from 1976 to 1980 had some type of birth defect. Miscarriages, cancer, and central nervous system problems were all found at high levels at Love Canal.

Luella Kenny lived in the northern end of Love Canal next to a creek. Her seven-year-old son Jon died and she believes his death was from dioxin poisoning. The creek where Jon often played was found to have 31 parts per billion of TCDD.

In 1980, a $15 million fund was established by the federal government to purchase all the homes in the neighbor-

tion of dioxin from the body is lactation. Breast milk is rich in fat and carries a considerable quantity of dioxin in that fat. Nursing a baby is one of the few things that speeds up the elimination of dioxin from the body, but this has the negative effect of exposing the baby to this toxic chemical. (See Chapter Eight, "Dioxin and Reproduction.")

Dioxin in the Liver

An important clue to how the body deals with dioxin—and in turn, how dioxin creates health problems—comes from the observation that at higher doses, a greater proportion of dioxin ends up in the liver than in fatty tissue. That is not what one would expect if the chemical just goes to the place where it dissolves best. Instead, it seems that higher amounts of dioxin *increase* the ability of the liver to retain dioxin. The best explanation available for this observation is that the liver, faced with high concentrations of this poorly soluble compound, makes more of the protein to which dioxin binds. The protein is called cytochrome P-4501A2.

Scientists have shown that this protein is quite abundant in the liver, and becomes more available in the presence of dioxin and related compounds. It actually does not bind to dioxin as tightly as does the Ah receptor (described later in this chapter), but it is the major protein to which dioxin binds because it is so abundant.

What Dioxin Does to the Body

The studies of dioxin and related compounds in animals make clear that these chemicals are toxic, and that their effects on health are very complex. It may at first seem strange that these compounds, which do not occur naturally, can interfere with normal

hood and to assist renters in relocating. Almost 900 families moved from the community, including 700 property owners and 200 families renting units in a housing complex.

Love Canal was "cleaned up" by building a trench around the canal and placing a clay cap over the top. Dioxin and the hundreds of other chemicals that leaked out of the canal still remain throughout the abandoned neighborhood. But that is not the end of the story.

The state of New York is now resettling parts of Love Canal. Families with limited options for owning a home of their own are moving into the northern end of the neighborhood, near the old homes where Luella Kenny and her neighbors once lived. Health authorities say that Love Canal is now habitable. What they will not say is that the neighborhood is safe.

processes in the body. Understanding how they do so is crucial to understanding the threat they pose, and perhaps figuring out how to prevent the problems they cause.

Dioxin affects health by interfering with the expression—the turning on and off—of genes in the body. In any individual, nearly every cell in the body contains the same genetic information. That genetic information resides in chromosomes, strands of DNA in the nucleus of the cell. The chromosomes in people contain more than 100,000 genes—discrete pieces of DNA that contain the directions for making proteins. These proteins are not only important building blocks of the body, but are also the agents of bodily functions. Since different cells have different functions, different cells need to use only some of their full set of genes.

Just as important, each cell needs to keep most of its genes turned off. For example, take the genes that control growth. In the embryo, cells divide and increase in number. When the individual's development is completed, the cells must stop dividing and start performing their specific functions. At this stage, normal development requires that the genes that control growth be turned off, and the genes that control particular cell functions—such as ovaries producing eggs or testes sperm—be turned on, or expressed. If these regulating functions are not successful, specific cells do not perform their designated role. For example, in some instances, genes that control growth would not "know" when to turn off, and cells would continue to increase in number, as they do in cancer.

Dioxin interferes with the cell's process of "knowing" when to turn certain genes off and on. Exactly how dioxin interferes is unclear at this time.

Under normal circumstances, genes function quickly, triggering a series of actions that determine normal cell function and then shut off. However, with dioxin in the receptor site, the message is different, and the normal process that determines specific cell functions is altered. With the presence of dioxin, gene function can either be blocked or kept continually on, as occurs with cancer.

How Dioxin Affects Cells

When dioxin enters a cell, it binds to a soluble protein called the "Ah receptor" (*a* for aryl and *h* for hydrocarbon) which is present in many parts of the body, including the liver, lungs, lymphocytes, and placenta. Once bound to the Ah receptor, dioxin can move about inside the cell. There it binds with a second protein called "Arnt" (the initials stand for Ah receptor nuclear translocator), which helps move the Ah receptor–dioxin complex into the nucleus of the cell. Once inside the nucleus, the dioxin-Ah-Arnt complex binds to DNA and turns genes on and off. (See Figure 4-1.) There is a long list of genes for which dioxin acts as a key, or switch; the one most commonly studied is called cytochrome P450A1 or CYP1A1. The genes that are turned on produce various proteins which influence hormone metabolism and growth factors and thus affect reproduction and immune system function. This complex can also cause genetic changes that result in cell proliferation, mutations, or cancer.

There are also other proteins involved in this binding process which can modify the complex. There appears to be a family of Arnt proteins that, when joined in a complex with dioxin and the Ah receptor, turn on different genes. This could account, in part, for the diversity of actions of dioxin in different tissues. In addition, these Arnt proteins and other involved proteins are defined by genes which can vary from individual to individual. This may explain why sensitivity to dioxin seems to vary widely among individuals. In some people, dioxin may cause chloracne, in others a suppression of the immune system, in others a change in hormone levels, and in still others all of these problems.

Dioxin is able to bind to the Ah receptor because it "looks like" (in size and shape) natural aromatic hydrocarbons that enter the body—for example, from plant foods like broccoli or cauliflower. However, there's a critical difference between dioxin and these naturally occurring compounds. Naturally occurring compounds are easily and rapidly metabolized in the liver. They are chemically modified by an enzyme called aryl hydrocarbon hydroxylase and then eliminated from the body. With dioxin, this detoxification system does not work well and as a result, dioxin remains unchanged in the body for long periods of time.

Thus, dioxin doesn't leave, as would natural hydrocarbons or normal hormones, which bind with the receptor, have a short-lasting effect, and move on. Instead, dioxin, comes in, binds to the receptor, and ties it up, preventing normal use of the site to regulate cell function. With dioxin bound to the receptor, a constant signal is sent from the nucleus to make more protein, simulating a switch stuck in the "on" position. This persistence of

---

Figure 4-1

Molecular "Cascade" of Action

Dioxin binds to the Ah receptor, which is about 100 times larger than dioxin. The small molecule that binds to the receptor is called a ligand. This receptor-dioxin complex binds to Arnt. The complex of dioxin + Ah receptor + Arnt moves into the nucleus and binds to DNA, where it can cause damage by turning genes on or off.

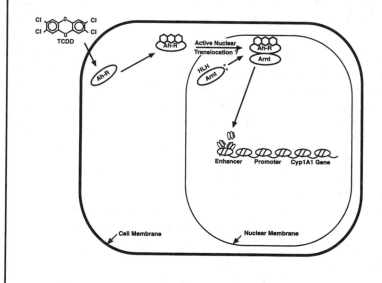

Source: USEPA, 1994

---

dioxin's actions may prove to be critical in understanding how it causes its many varied effects.

The fact that dioxin binds to the Ah receptor is clearly central to our response to dioxin. The Ah receptor binds to dioxin at the concentrations at which dioxin has its biological effects. In addition, the more toxic a member of the dioxin family is, the more tightly it binds to the Ah receptor. And there are genetic variants of the Ah receptor itself, and differences in those receptors correlate with different responses to dioxin.

Why is it significant that dioxin acts through binding to a receptor? First, the fact that dioxin binds to a receptor explains why it is such a potent biological molecule even at such low concentrations.

Second, the EPA report and other studies disprove two theories proposed by dioxin-producing–industries. The first is that not all toxic effects of dioxin relate to its relationship with the Ah receptor; the second is that dioxin must bind to a minimum number of receptors for any effect to occur, and that exposure to dioxin at levels below this "threshold" will not cause any adverse health effects. Both of these theories have been seriously damaged by evidence pulled together in the EPA report. The EPA identified a substantial body of evidence from animal studies indicating that the Ah receptor mediates the biological effects of dioxin. And, although the studies of human tissue are much less extensive, it seems reasonable to assume that dioxin's effects in humans are also Ah receptor–mediated. Based on all the data reviewed in the EPA report, the agency commented that "the Ah receptor presumably participates in every biological response to TCDD."

Scientists now understand to a considerable extent the mechanism by which dioxin behaves in mice. The story of dioxin's behavior in human beings is far less complete, but all the evidence suggests that mice and humans respond to dioxin in similar ways. The Ah receptor in mice is very similar to the Ah receptor in humans, so laboratory studies on dioxin in mice may reflect accurately what happens in humans. Recent research findings provide further opportunity to learn about dioxin's effect in people. In May 1995, researchers at the National Cancer Institute produced mice that lacked the Ah receptor (Fernandez-Salguero, 1995; Stone, 1995). Because the Ah receptor in humans is very similar to the Ah receptor in mice, these "knock-out" mice present the opportunity to learn a great deal about the role of the Ah receptor in causing the toxic effects of dioxin. It is now possible to determine whether any of dioxin's effects do in fact occur independently of the receptor. Using these mice, scientists will be able to compare dioxin's health effects in knock-out mice to those in normal mice who have the Ah receptor.

The knock-out mice were produced by using genetic techniques to "knock out" the gene that makes the Ah receptor protein. These knock-out mice also showed signs of immune system and liver damage which suggest that the Ah receptor is vital to the liver and immune system development. (See Chapter Six, "Dioxin and the Immune System.")

Contributors

Dr. Frank Solomon
Dr. Beverly Paigen
Stephen Lester, M.S.

Chapter Five

How Much Dioxin Are We Getting?

How much dioxin and dioxin-like chemicals are we getting in our bodies? At what levels do they cause harm? The simple answers are these: We are getting too much, and the levels we now have in our bodies are at or close to the levels that cause harm.

There are two different ways to evaluate dioxin exposure. The first is by the amount of dioxin we're exposed to daily. This value helps assess our risk of cancer. The second is by the total amount now in our bodies. Because dioxin is so stable and has such a long half-life, we accumulate dioxin over our lifetimes. The total amount now in our bodies, called the *body burden*, helps assess the risk of health problems such as endometriosis, suppression of the immune system, disruption of hormone activity, and reduced sperm count.

A "Safe" Daily Dose of Dioxin

In order to evaluate public health risks, the EPA has determined that a "virtually safe dose" of dioxin is 0.006 picograms per kilogram body weight per day (pg/kg/day) or 0.42 pg/day for a 150 pound adult. According to EPA, this dose will cause only one

Picograms to Nanograms

A nanogram is 1,000 times larger than a picogram. How does the intake of dioxin, which is measured in picograms, turn into a body burden level of dioxin that is measured in nanograms? Dioxin is broken down very slowly. So we ingest new dioxin more quickly than we break down what's already in our bodies. We accumulate more and more dioxin, and with the passage of time, the picograms we eat in our food build up into nanograms stored in our bodies.

additional cancer for every 1 million exposed people, if that exposure occurs over a seventy-year lifetime.

Based on this "safe" dose, everyone is consuming too much dioxin. The EPA estimates the average intake of dioxin for the general population is 119 pg/day. This is more than 280 times the EPA's safe dose of 0.42 pg/day. This value poses a cancer risk in the range of 1 in 10,000 to 1 in 1,000, which is 100 and 1,000 times higher, respectively, than the normal "acceptable" risk level of 1 in 1 million. This is one of the highest cancer risk levels ever reported. Other, non-cancer, effects of dioxin are not considered in this risk calculation.

A "Safe" Body Burden of Dioxin

Because dioxin is so stable and has such a long half-life (seven to eleven years) in the body, we accumulate dioxin over our lifetimes. The EPA calculated the average level of dioxin in the body to be 9 nanogram per kilogram (ng/kg), or 9 parts per trillion (ppt). (Also see conversion chart in Appendix B.) This is an average level for a middle-aged person. Some people will have higher levels and some lower; in general, younger people will have lower levels and older people will have higher levels.

Using a number of studies in both animals and humans (many of which are reviewed in subsequent chapters), the EPA

Table 5-1

Levels of Dioxin Known to Cause Health Problems

Body Burden (ng/kg)	Species	Health Effect
7	Mice	Increased susceptibility to viruses
7	Monkeys	Altered immune response
14	Humans	Altered glucose tolerance
14	Humans	Decreased testes size
19	Monkeys	Learning disability
54	Monkeys	Endometriosis
64	Rats	Decreased sperm count
83	Humans	Decreased testosterone

Source: USEPA, 1994b

calculated that the lowest body burden level that has been shown to cause harm in people is 14 ng/kg. This level is just above the average American's body burden of dioxin of 9 ng/kg. This fact is the basis of the statement, made several times in the EPA report, that "levels of dioxins are at or near the levels known to cause harm." These numbers are as disturbing and as frightening as the cancer risk assessment numbers.

If the average person has a dioxin body burden levels of 9 ng/kg, and depression of the immune system occurs at 7 ng/kg (see Table 5-1), then the body burden of the average person is not "near or at," but actually *above* the level of dioxin that has been shown to cause harm. Any further increase in dioxin will certainly be harmful.

The information in Table 5-1 immediately counters two claims made by the chemical industry and others who are trying to downplay the dioxin problem. One claim is that different species respond very differently to dioxin, and humans are among

The Columbus Trash Burning Power Plant

We do not fear our enemies more than we love our children.

For eleven years, the Columbus trash plant belched black smoke over the capital of the State of Ohio. Little did anyone know what else was coming out of the stacks. Not until 1993, when citizens started asking questions and doing file reviews at the Ohio Environmental Protection Agency, was it revealed (by the citizens) that the plant had been tested for dioxin three times. All of the data showed extremely high levels of dioxin. The local grassroots group PARTA (Parkridge Area Residents Take Action) discovered that in 1992 a stack test revealed dioxin emissions amounting to 1 kilogram of dioxin TEQ being released per year from just one of the plant's six stacks.

After filing complaints with the Ohio EPA and the U.S. EPA, the group held meetings with local citizens as well as state and federal elected officials. PARTA came to realize that the system was taking too long to take action on the trash plant. We wanted the plant closed

the least sensitive. Yet, as shown in this table, several of the lowest levels observed to cause health effects come from human studies. Another claim by industry is that there is a "threshold" for dioxin exposure below which no health effects will occur. (See Chapter One, "The Toxicity of Dioxin.") But when the average American already carries a dioxin body burden that is higher than a level known to cause harm, the concept of threshold is irrelevant. Even if there were a safe threshold, we are already past it. For the same reason, the use of risk assessments to evaluate health risks has no relevance because this process assumes a starting point of zero, a point we have already passed.

The body burden of the average person tells only part of the story. Some people's body burden will be above the average, and it is important to know how many are above and how high their dioxin levels are. This is not easy to determine from the existing scientific literature. Many studies used "pooled" samples of large numbers of people so that individual differences are obscured. For the few studies that did measure individuals, the distribution of dioxin levels in the body shows that there are substantial numbers of individuals at the higher levels. According to the EPA, this distribution suggests that about 1 percent of the population has a body burden that is 7 times the average.

One survey of body burden, the National Human Adipose [fat] Tissue Survey (NHATS), conducted by the EPA, used samples from 865 individuals mixed in pools of about 20 each (Orband, 1994). The national average dioxin level found was 28 pg TEQ per gram of lipid (blood fat) or 28 ppt. These levels increased with age, but did not vary with race, sex, or geographical region of the U.S. This survey was done in 1982 and repeated in 1987. The 1987 survey showed some decreases in dioxin body burden. However, with only two points in time studied, it is not possible to judge whether dioxin is truly declining or whether some problems in the study resulted in a false decline. Up until a few years ago, the only

and we would not take no for an answer. As part of our strategy, PARTA decided to contact William Sanjour of the Solid Waste Division of the U.S. EPA. After a long and sometimes heated discussion, Sanjour was convinced to write a memo to EPA administrator Carol Browner about the dioxin emissions from the plant. On February 12, 1994, the day Mr. Sanjour's memo hit the local media, all hell broke loose. It was great! Every one of the local elected officials wrote to Browner demanding an explanation for Sanjour's memo. As of this writing, Browner has yet to offer any explanations to anyone, because she knows that everything in Sanjour's memo was correct.

It might be hard for most people to believe, but the U.S. EPA was there for us. Of course it would have been very hard for the EPA to explain why it was proposing a 30 nanogram limit for dioxin for new incinerators, while letting the Columbus plant spew 596 times this amount of dioxin over the capital of Ohio. The EPA knew that if it really believed what its own dioxin reassessment revealed, then it had to move on Columbus. In August 1994, the U.S. EPA filed a unilateral order

> against the Columbus trash burning power plant, requiring the plant to be retrofitted with new pollution control equipment within two years. However, the cost of the retrofit was too great, and on November 2, 1994, a unanimous vote by the Solid Waste Authority of Central Ohio closed what was possibly the largest known single source of dioxin in this country.
>
> The people of Central Ohio can breathe easier for a while—that is, until the plant is sold to another company that wants to burn trash. Will dioxin exposure ever end? You're damn right it will, if we strive to create a unified force from the varied voices demanding a healthy and safe environment for all. Our children deserve our best efforts on their behalf.
>
> —*Teresa Mills, PARTA*

way to estimate the body burden of dioxin was to take samples of fat tissue, which are not easy to obtain for large studies. Then scientists discovered that dioxin was carried in the fat, or lipid, part of the blood, and that dioxin in these blood lipids was in equilibrium (balance) with dioxin in fat tissue (Papke, 1989; Schecter, 1990). This was an important finding because blood is easier to obtain from people than fat. However, it still is not easy for any individual to find out his or her own body burden, because measurements cost between $1,500 and $2,500, and because there is a scarcity of laboratories equipped to carry out the measurements (Schecter, 1994a). In contrast, the body burden levels of PCBs, a dioxin-like substance, are relatively easy to measure. This costs about $75 to $125 per sample, and many laboratories can do the measurement (Schecter, 1994b).

Highly Exposed Groups

Most human dioxin exposure is through food. Since food found in the local supermarket comes from all over the United States, dioxin exposure averages out for people whose food comes off grocery store shelves. However, if a person eats a

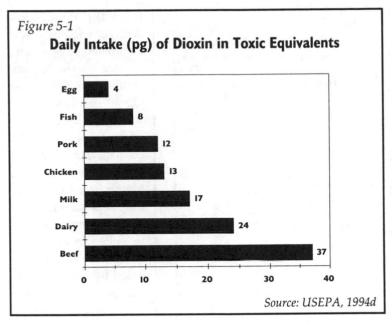

Figure 5-1
Daily Intake (pg) of Dioxin in Toxic Equivalents

Source: USEPA, 1994d

great deal of local food, and if that local food is heavily contaminated with dioxin, that person can accumulate an extra high dioxin concentration. This does not apply to fruits and vegetables from your own garden, even if you have a large garbage incinerator or copper smelter in your city because of the poor uptake of dioxin by plants and vegetables. Fruits and vegetables contain negligible amounts of dioxin. It does apply to food higher up the food chain which is produced near a local dioxin source. For example, if you live on a dairy farm and consume your own milk and beef, and if there is a large source of dioxin nearby, then you are likely to be among the highly exposed.

Another highly exposed group is people who consume a lot of freshwater or coastal fish. Some people rely on fish as a main staple of their diet. If these fish are caught in lakes, rivers, or streams that are contaminated with dioxin, these people risk high levels of dioxin toxicity. This is especially true for rivers that are downstream from a paper or pulp mill. In Maine, for example, fisherman are advised to eat no more that two fish meals a year

Table 5-2

Dioxin in U.S. Foods

Food type	Total TEQ (pg/gram food)
Ground beef	1.5
Soft blue cheese	0.7
Beef rib steak	0.65
Lamb sirloin	0.4
Heavy cream	0.4
Soft cream cheese	0.3
American cheese stick	0.3
Pork chops	0.3
Bologna	0.12
Cottage cheese	0.04
Beef rib/sirloin tip	0.04
Chicken drumstick	0.03
Haddock	0.03
Cooked ham	0.03
Perch	0.023
Cod	0.023

This table shows the average level of dioxin found in different foods. By multiplying the average dioxin level in food by the average consumption rate of these foods, as shown in Table 5-3, you can estimate the level of dioxin consumed when eating these foods.

Source: Schecter, 1994c

from the Penobscot, Androscoggin, and Kennebec Rivers, due to high dioxin concentrations (MDEP, 1994). Ocean fish have lower levels of dioxin, but coastal fish, mussels, clams, oysters, and bottom-feeding fish have higher levels of dioxin due to pollution of the coastal sediments.

People who live in a community contaminated by a local dioxin source are at even higher risk of suffering one of the many health problems caused by dioxin. People living in dioxin-contaminated communities such as Times Beach, Missouri; Jacksonville, Arkansas; and Pensacola, Florida; are exposed to dioxin leaking from contaminated sites located in these communities. The exposures that result from living near dioxin-contaminated sites add to the general exposure that these people receive from food.

Table 5-3
U.S. Consumption Rate for Various Food Groups

Food group	Consumption rate (gm/day)	Low range of dioxin (pg/gm)	High range of dioxin (pg/gm)
Fruits and vegetables	283		
Milk	254	0.04	0.04
Beef	88	0.04	1.50
Other dairy products	55	0.04	0.70
Poultry	31	0.03	0.03
Pork	28	0.03	0.30
Fish	18	0.02	0.13

Source: Schecter, 1994c

Sources of the Daily Dose of Dioxin

The family of dioxin-like chemicals enters the body almost entirely through the food chain. The EPA report concludes that 90 percent of the dioxin people are exposed to comes from the food they eat. In trying to assess the risk of consuming dioxin in food, the EPA determined the average "daily intake" of dioxin. Since dioxin and dioxin-like chemicals are contained in fat (and, to some extent, in liver), we are exposed primarily through ingestion of fat in meat, milk, cheese, eggs, and fish. Two different estimates of daily intake have been made, which agree reasonably well with one another. The first is the intake estimated by the EPA, shown in Figure 5-1.

The EPA estimated that the average 150-pound adult ingests 119 picograms of dioxin per day. About 95 percent of this comes from beef, chicken, pork, fish, and dairy products such as cheese, milk, and eggs. Vegetables, fruit, grains, and water contain negligible amounts of dioxin.

Table 5-4

Dioxin Equivalents Consumed Each Day at Different Ages

Age	pg/kg/day
Breast-fed infant	34–53
Formula-fed infant	0.07–0.16
1–4	1–32
5–9	1–27
10–14	0.7–16
15–19	0.4–11
Adult >20	0.3–3

Source: Schecter, 1994c

The EPA's values for average dioxin intake are based on the average dioxin levels in food multiplied by the average food consumption. For example, the average daily intake of dioxins through fish is low because average fish consumption in the U.S. is low. However, averages can be misleading. For example, some freshwater and coastal fish contain much higher levels of dioxin, especially when caught downriver from a dioxin source. In Maine, fish caught downstream from a pulp and paper mill had between 2 and 3 pg/gm dioxin (MDEP, 1994). This level is consistently higher than the average dioxin level found in ground beef, as shown in Table 5-2.

The second estimate of dioxin levels in U.S. foods comes from the work of Dr. Arnold Schecter of the Department of Preventive Medicine at the State University of New York at Binghamton. Dr. Schecter conducted one of the few surveys in the U.S. of dioxin levels in foods taken off supermarket shelves (Schecter, 1994c). Based on this work, Schecter calculated that the range of dioxin intake for an adult to be between 0.3 and 3 pg/kg/day. To compare this estimate to the EPA's estimate, multiply each number by 65 (the size of the "standard" person; equal to 143 pounds) to obtain 18 to 192 pg/day—in good agreement with the EPA's average estimate of 119.

These estimates of the daily intake of dioxin in the United States are similar to estimates made for the Swedes, the Dutch, the English, and everyone else whose diet has been tested for dioxin (Beck, 1994; Patterson, 1994; Schecter, 1994). Even groups who live in remote places take in dioxin. The Inuit people of northern Canada have a high intake of dioxin and PCBs due to the fact that their primary diet is composed of fish, seal, and whale, all animals high in the food chain and high in fat (Dewailly, 1994).

Schecter also examined typical food consumption at various ages and calculated the daily consumption of dioxin for different age groups, as shown in Table 5-4.

All of these values substantially exceed the EPA "safe" dose of 0.006 pg/kg/day, which is very disturbing. The average daily intake of dioxin in breast milk is at least ten times higher than the average daily intake for an adult. More alarming, however, is the fact that after one year of nursing the daily intake for breast-fed infants is ten times the EPA "safe" dose, which is based on a lifetime of exposure. Because people are at the top of the food chain, the dioxin content in breast milk is higher than in any other food. Breast feeding has wonderful health benefits, as well as psychological benefits, to baby and mother. But breast milk also has dioxin, posing an agonizing choice for each nursing mother. This is a choice no parent should have to face, and steps must be taken to stop dioxin exposure and reduce the levels of dioxin in breast milk. (Breast milk and dioxin is discussed in greater detail in Chapter 8, "Dioxin and Reproduction.")

Dioxin is a very serious environmental problem. The solutions are both personal and political. On an individual level, it will help to eat less fat of animal origin. Skim milk is better than whole, lean meat better than fatty, skinless chicken better than whole. But there is no way for us, as individuals, to completely eliminate dioxin from our food and our lives. Steps must be taken to stop dioxin exposure by stopping the production and release of dioxin. No one has the right to poison our food supply.

Contributors

Dr. Beverly Paigen
Stephen Lester, M.S.
Teresa Mills

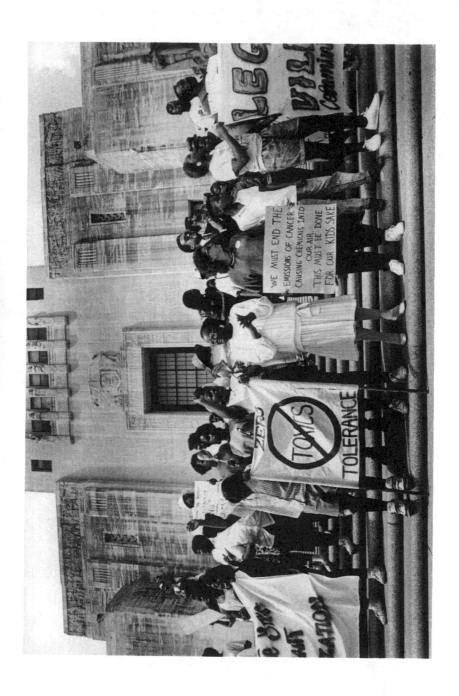

Second Louisiana March Against Poison, April 22, 1990. © *Sam Kittner*

Chapter Six

Dioxin and the Immune System

The immune system is a network of cells that detects and eliminates "foreign invaders" such as bacteria, viruses, parasites, and cancer cells. In addition to recognizing foreign and harmful cells, the immune system must recognize and avoid reacting against "self," or normal, cells in the body. Mistakes in either direction cause health problems.

If the immune system is suppressed, one may be more susceptible to infectious diseases and to cancer. An overreactive immune system can lead to allergies and autoimmune diseases. In allergy, the immune system correctly recognizes something as foreign, such as ragweed pollen, but incorrectly reacts against it as if it were harmful. In autoimmune diseases such as lupus and scleroderma, the immune system mistakes the body's own normal cells for foreign cells, and attacks them.

The immune system is very complex, composed of many different cell types that interact in highly interdependent and poorly understood ways. Proper functioning of the immune system depends upon delicate control over the growth and maturation of immune system cells and the hormone-like compounds that regulate them. Dioxin can be a powerful biological dysregulator, upsetting the proper functioning of a number of the body's physiological systems, including the immune system.

East Liverpool, Ohio

For fourteen years, we have been fighting to prevent the construction and operation of the world's largest commercial toxic waste incinerator in East Liverpool, Ohio. The Waste Technologies Incorporated (WTI) incinerator, owned by Von Roll, a Swiss company, is located on the banks of the Ohio River in a residential neighborhood. The incinerator stack is level with a 400-student elementary school that sits on a bluff less than 1,100 feet away. East Liverpool is an economically depressed community of conservative, hardworking essentially, blue-collar people, poor people, and people of color. All of the town's 500 African American residents live in the vicinity of the incinerator.

Both Bill Clinton and Al Gore have expressed concern over WTI's location. Al Gore called WTI an "unbelievable idea," saying, "A lot of people don't know how high the stakes are." Even the EPA regional director who permitted the incinerator, Valdas Adamkus, admitted that WTI should never have been built where it is!

One central organ in the immune system is the thymus gland, located behind the breast bone. The thymus is relatively large at birth and during childhood, but it shrinks with age. Certain key cell types that originate in the thymus are called "T cells." T cells are "educated" in the thymus to be able to distinguish between self and foreign invaders. This "education" occurs in the fetus and in early infancy, as the immune system matures.

One of the first observed toxic effects of dioxin was a dramatic shrinking of the thymus in young animals. Because the thymus is important to the immune system during early life, it is possible that the fetus or infant is especially sensitive to the immunotoxic effects of dioxin. Scientific studies show that animals exposed before birth are very susceptible to immune suppression caused by dioxin and dioxin-like compounds.

The immune system is very sensitive to the effects of dioxin. Suppression of the immune system takes place at doses much lower than those which cause the visible shrinking of the thymus. Dioxin decreases the resistance of mice to bacterial, viral, and parasitic infections and to cancers. Mice infected with influenza virus died at a higher rate if they were first exposed to a single dose of dioxin at 10 nanograms/kilogram of body weight. This is the smallest single dose of dioxin that has been observed to have a toxic effect.

Recently, there have been a number of mass deaths of marine mammals. It has been suggested that these deaths were caused by viral disease in animals with immune systems suppressed by pollutants. Recent experiments demonstrated that captive seals who were fed fish from the heavily contaminated Baltic Sea had a weaker immune response than seals who were fed fish from the cleaner Atlantic Ocean. The levels of dioxin-like compounds, particularly PCBs, were several times higher in the seals fed Baltic herring (Ross, 1995).

These studies are important because, like people, the seals were exposed to many of the dioxins and dioxin-like chemicals that are typically found in the environment, not to a defined dose of laboratory TCDD. The seal study adds fur-

This isn't a tragedy in the making; the tragedy has already begun. I initially became concerned when I learned that WTI would be permitted to emit 4.7 tons of lead every year. Any amount of lead released into the environment is cause for concern, but it becomes criminal when you consider that it will be raining down on 400 schoolchildren next door. In a state-funded health study of East Liverpool schoolchildren, 69 percent of those tested had no mercury in their urine in March 1993. By September 1993, after WTI dumped twenty-nine pounds of mercury into the air during two days of a test burn, and then continued to operate for six months, almost the same percentage of children now *did* have mercury in their urine!

The EPA is currently conducting an indirect food chain risk assessment for WTI. To many, that would appear to be an acknowledgment that EPA doesn't know what the risk is. And yet the WTI incinerator continues burning. Our children are the "mine canaries."

During the WTI federal court case in February 1993,

the "Guimond memo" surfaced. Sent to EPA chief Carol Browner on January 22 by Richard Guimond, the senior U.S. public health service officer in the EPA, the memo details a strategy to avoid acknowledging the serious health threat from dioxin emissions at WTI, and describes how the EPA can explain to the court its failure to assess food-chain exposures at WTI. Guimond also writes that risks due to consumption of dioxin contaminated beef in the vicinity of WTI can be 1,000 times greater those considered in the inhalation-only risk assessment. Based on this thousand-fold risk increase and the EPA's prior finding that inhalation of dioxin alone would pose a cancer risk of 1.3 per million, the cancer risk posed by dioxin would be 1300 times greater than the EPA "acceptable risk" standard. That's 1,300 additional cancers from dioxin alone!

Making a December 7, 1992, post-election promise, President-elect Clinton and Vice President-elect Gore, in a prepared press release, said that WTI should not operate until questions of safety

ther support to the finding that the levels of dioxin and dioxin-like compounds found in the environment may be high enough to suppress the immune system. However, it is possible that other Baltic Sea pollutants may have contributed to this effect as well.

How Dioxin Affects the Immune System

Despite many years of research, the mechanism by which dioxin affects the immune system is not precisely known. Dioxin is thought to affect individual cells by first reacting with the Ah receptor. (See Chapter Four, "Dioxin Inside Our Bodies.") Dioxin's immunotoxic effects appear to extend beyond the impairment of individual cells or cell types. It also affects the interaction of the complex network of immune cells. Thus, the toxic effects of dioxin in an animal are often greater than those predicted from dioxin's effects on immune system cells isolated in a test tube. This suggests that "indirect" effects on the immune system may be quite important. Dioxin may act in part through hormonal systems, or by altering the behavior of cells that nurture and support the primary cells of the immune system.

While most of dioxin's toxic effects are thought to hinge on its binding to the Ah receptor, it is not known for certain whether this is true for its effect on the immune system. As described in Chapter Four, scientists at the National Cancer Institute have provided new evidence linking the Ah receptor protein and the immune system (Fernandez-Salguero, 1995). These researchers "knocked out" the Ah receptor gene in mice. Because this gene leads to the detoxification of many chemicals like dioxin, these researchers at first anticipated that mice without the Ah receptor would be useful for studying the role of the Ah receptor in determining the toxicity of dioxin and other environmental chemicals. To their surprise, about half the mice without Ah receptors died within the first few weeks of life from infectious disease. Although the precise cause of death was not known, a poorly functioning immune system

and legality were answered. Those questions have now been answered. Using the EPA's own risk criteria, a federal court ultimately ruled that even one year of WTI operation posed a substantial and unacceptable risk. Ohio Attorney General Lee Fisher has made a legal conclusion that WTI was built and is operating the incinerator without a permit, in felony violation of the law.

Yet Carol Browner's EPA has worked aggressively to remove any and all obstacles from WTI's path, becoming the first EPA ever to allow a commercial hazardous waste incinerator to proceed to commercial operation immediately after failing its test burn. Ms. Browner's EPA has also ignored substantive facts confirmed by the Ohio attorney general and the federal court.

An explanation for the administration's retreat on the WTI issue may lie with the fact that a large portion of the project was bankrolled by the Stephens family, wealthy investment bankers from Arkansas. The Stephens family has ties to the Clinton campaign, contributing over

$100,000 to help Clinton's presidential election bid in 1992.

WTI is the epitome of everything that is wrong with the environmental regulatory process. Little in the way of facts, sound argument, or documented violation of law has had any effect. The EPA has tried to manipulate science to provide the illusion that WTI is safe. Because of the financial interest behind WTI, politics has suppressed science.

—Terri Swearingen
(Orange Resource Book, 1995)

that failed to protect the mice from infection may have been responsible. These mice had very low numbers of lymphocytes (white blood cells)—key immune cells—which indicates a suppressed immune system. By feeding the mice sterilized food and keeping them in a special germ-free environment, researchers were able to keep some mice alive. These mice had slightly higher numbers of lymphocytes than the mice that died, but their lymphocyte count decreased upon reaching adulthood. These mice also had abnormal livers with extensive scar tissue.

This experiment suggests that the Ah receptor is essential for normal development and functioning of the liver and immune system. It may be that dioxin effectively takes the Ah receptor out of circulation, so that it cannot do its normal job in the immune system. This would result in health effects similar to those found in mice with no Ah receptor. Further studies like these may tell us why dioxin is so immunotoxic.

Dioxin and Human Immune Function

Can these results in mice be extrapolated to people? Studies show that the Ah receptor in mice is very similar to the Ah receptor in humans, that mice and humans respond similarly to the toxic effects of dioxin, and that mice do provide a good model for predicting dioxin's effects on people (Birnbaum, 1994). Based on the available evidence, it is prudent to assume that people may be as susceptible as mice to the immunotoxic effects of dioxin.

Information on dioxin's immunotoxicity to humans is limited and sometimes contradictory or inconsistent. Studies of accidental or occupational exposures to dioxins, furans, or PCBs have generally shown increased frequency of infections, especially of the skin and respiratory system. Tests on the exposed individuals indicated some impairment of immune system function. Such immune system damage was observed in residents of the Quail Run Mobile Home Park in Times Beach, Missouri, a notorious dioxin-contaminated site (Hoffman, 1986). However, subsequent investigation of the same group failed to confirm the initial finding, although a shift in a particular protein (thymosin alpha-1 levels) was related to the dioxin levels in the subjects' fat tissues (Evans, 1988). The practical meaning of this finding is not clear.

Immune system toxicity was also studied in the children exposed to dioxin in the Seveso, Italy, explosion (Reggiani, 1978). One study showed a decrease in complement, a set of proteins that take part in the immune response, but another study failed to find any differences between children in the exposed group and the control group for the small number of immune parameters investigated.

One possible reason that studies of immune function in humans are so limited and contradictory is that we do not have good tests for measuring how the various parts of the human immune system interact. This interaction is thought to be one of the main ways dioxin affects immune functioning. If one were to take snapshots of any complicated machine with interacting moving parts, such as a car engine, one would see engines at all different stages of the cycle. Some snapshots would have pistons up and some would have pistons down, but the photographs wouldn't necessarily help one understand how the parts interact with each other or how the engine works. Similarly, by only looking at "snapshots" of immune mechanisms at certain points in time, we miss the way the parts connect and relate to each other. In addition, the EPA report described many problems and limitations in the design of many of the studies that have looked at immune system damage in groups of exposed people. A common problem has been determining who was exposed and at what levels. As a result, many of the studies classify people with little or no dioxin exposure as "exposed," thus diluting comparisons to truly exposed people.

Despite the problems in our understanding of the immune system, the EPA report describes evidence that is consistent with immune effects caused by exposure to dioxin at levels currently found in the environment.

Contributors

Thomas Webster
Dr. David Ozonoff
Dr. Beverly Paigen
Stephen Lester, M.S.
Terri Swearingen

Chapter Seven

Dioxin and Cancer

Dioxin causes cancer in many different species of animals, including humans. Dioxin causes cancer in many different parts of the body. Dioxin causes cancer in both males and females. So in scientific jargon, dioxin is a "multi-species, multi-site, multi-sex carcinogen" (cancer-causing chemical). This is about the strongest statement that science can make about the potential for a chemical to cause cancer.

Dioxin does not cause cancer in the traditional way, by changing or mutating DNA. Chemicals that change DNA are called "direct" carcinogens. Short-term tests (short-term because they provide answers in a matter of days) that predict whether a chemical will cause cancer depend on the chemical's ability to mutate DNA. Since dioxin does not mutate DNA, it is "negative" in most of these tests. Instead, dioxin is an "indirect" carcinogen. Some other chemical or radiation causes a change in DNA, and an indirect carcinogen (dioxin) increases the probability that this change in DNA develops into a full-blown cancer. An indirect carcinogen, sometimes called a "promoter," can work by accelerating cell growth, by suppressing the immune system, or by affecting hormone function.

A chemical that causes cells to grow or causes some growth factor to increase can lead to cancer. Cell mutations that lead to cancer occur with some regularity in the body, but to actually turn into cancer, a cell needs to divide. A skin cell, for example, might contain a mutation that could lead to cancer, but that skin cell may never divide, and thus never pose a problem. The cell simply wears off eventually or dies, and the potentially cancerous mutation it contains disappears with it. However, if there was a stimulus for that cell to divide, caused by exposure to a chemical like dioxin, the mutation capable of causing cancer would be expressed, and skin cancer would result. Dioxin increases the activity of epidermal growth factor (EGF) which leads to increased cell division. This is one mechanism by which dioxin could lead to the development of cancer.

As described in Chapter Six, "Dioxin and the Immune System," a chemical that suppresses the immune system can also lead to an increase in cancer. The job of the immune system, sometimes called "immunosurveillance," is to watch for anything foreign in the body such as bacteria, viruses, or cancer cells. If a foreign substance is found, then the immune system begins a set of events to get rid of the foreign substance. When cancer cells arise in the body, most of the time they are detected and destroyed by the immune system. But if the immune system is not working properly, then the cancer cell may escape surveillance and keep growing and dividing until it is too big for the immune system to handle. Since dioxin suppresses the immune system, this could be another mechanism by which dioxin leads to the development of cancer.

In addition, dioxin increases the production of certain proteins in the body (See Chapter Four, "Dioxin Inside Our Bodies.") The proteins, called cytochrome P4501A1 and cytochrome P4501A2, lead to the activation of chemicals that bind to DNA and cause mutations. This could be another mechanism by which dioxin leads to the development of cancer.

And finally, several major cancers are hormone-sensitive, such as breast, ovary, testes, and prostate cancer. If a change in the hormonal balance in the body takes place, then hormone-sensitive cancers can grow more rapidly. Dioxin acts as a weak hormone in some cases and interferes with normal hormonal action in other cases. Dioxin's effect on hormones is another mechanism by which it could lead to the development of cancer.

Studies of Dioxin and Cancer

Laboratory Studies

Four long-term studies using large numbers of animals have shown that dioxin causes cancer in animals (Kociba, 1978; NTP, 1982; Della Porta, 1987; Roa, 1988). These studies, involving mice, rats, and hamsters, show that dioxin (specifically TCDD) causes cancers in the liver, lung, tongue, roof of the mouth, nose, thyroid gland, thymus gland, adrenal gland, skin of the face, and under the skin. Overall, these animal studies prove that dioxin is a multi-site carcinogen in several species.

The Limitations of Human Studies

Human studies are more difficult to control than animal studies. One reason is that the scientist is dependent upon accidental, workplace, or environmental exposure for data. Exposures in these settings are rarely measured, and seldom defined or well understood. In fact, these exposures very often involve multiple chemicals. Linking human exposures to disease is the science of epidemiology—the study of whether a suspected cause is linked to a disease in humans.

Epidemiology yields the simplest and most accurate results either when an exposure produces one disease, or when a disease has only one cause and appears soon after exposure. Unfortunately, cancer epidemiology benefits from none of these qualities. Exposure to cigarette smoke, for example, may cause several forms of cancer: cancer of the lung, cancer of the bladder, cancer of the pancreas, and perhaps even leukemia. And none of these cancers is caused only by cigarette smoking. To make it even more confusing, not all smokers develop even one of these diseases. And all these cancers take years to appear after the initial exposure. Because of this time lag, good measurements of exposure are rarely available, making the connection between cause (exposure) and effect (disease) even harder to establish.

The body's response to dioxin exposure should be related to the level of exposure. Therefore, a study of people who are likely to have had high exposures (industrial workers, for example) contrasted with a study of people who have had little or no exposure could help uncover the effects of exposure. Thus when trying to determine if a chemical causes cancer, the group of people first studied are often workers who must handle or process the chemicals on their jobs, or people exposed as a result of a large-scale accident, such as the residents of Seveso, Italy.

The less accurate the measurement of exposure is, the harder it is to show a relationship between exposure and disease. Determining the actual amount of exposure (or dose) is the most difficult task in cancer epidemiology. This is especially true for dioxin studies because, until recently, actual measurements of the amount of dioxin to which people were exposed were not made. Instead, two rather crude, indirect measures of exposure were used: the number of years a person worked at a job where exposures to dioxin occurred, or the presence of chloracne, a severe skin condition caused by dioxin.

Many things can happen that can confound or confuse epidemiologic studies. The estimate or measurement of exposure can be inaccurate. Other causes of the same disease can be present. The control or comparison populations may not be truly comparable to the exposed populations. People may not remember, or know, what they have been exposed to, or they may "remember" exposures that never really happened. However, if the relative risk of disease—determined by comparing the rate of disease in exposed people with the rate of disease in unexposed people—is high, it is less likely that errors or biases occurred in the study.

Studies on Human Exposures

Human studies of dioxin and cancer are usually described as "relative risks with a 95 percent confidence range." What does this mean? A relative risk of 1.0 means that an exposed person is not at any increased cancer risk compared with an unexposed person. A relative risk of 2.0 means that an exposed person is twice as likely to get cancer as an unexposed person. The 95 percent confidence range is the result of a statistical test. A relative risk of 4.0 with a 95 percent confidence range of 3.0 to 5.0 is a very significant risk, and cannot be explained away as a chance occurrence. However, the same relative risk of 4.0 with a 95 percent confidence range of 0.5 to 7.5 is not very convincing, because the confidence range includes a risk of 1.0, which means there is no increased risk. Therefore, a relative risk with a confidence range that includes 1.0 is said to be non-significant: The result could have occurred by chance.

There are many possible reasons for a wide confidence range. One of the most common reasons is that the number of people studied was small. The smaller the number of people in the study, the wider the range. Conversely, the larger the number of people in the study, the smaller the 95 percent confidence range, and the more likely it is that any observed risk is real. A well-designed study will have a small confidence range.

There have been many studies of workers exposed to dioxin. Many of the older studies are not very useful because no information on the levels of exposure to dioxin were available. However, there are three studies of high quality with good exposure data. These provide the basis for concluding that dioxin causes cancer in people. The first is a study conducted by the National Institute for Occupational Safety and Health (NIOSH). In this study, 5,172 workers at twelve plants in the United States, all of which produced chemicals contaminated with dioxin, were tracked for over twenty years. Men exposed for over one year had a 50 percent increase in stomach cancer, lung cancer, non-Hodgkin's lymphoma, Hodgkin's disease, and cancer of soft and connective tissues (soft tissue sarcoma). The relative risk for these cancers was 1.46, with a 95 percent range of 1.2 to 1.8. The largest relative risk was 9.2 (95 percent range 1.9 and 27.0) for connective and soft tissue cancers. This is more than nine times the expected cancer rate. The excess lung cancers could not be explained by excess cigarette smoking, as other smoking-related deaths did not increase (Fingerhut, 1991).

The second critical study forming the basis for the conclusion that dioxin causes cancer in people is a study of German workers. This study examined 1,583 workers employed in a BASF plant that produced herbicides contaminated with dioxin. These workers were found to have an increased risk for all cancers, which increased with the duration and intensity of exposure. Exposure levels were determined by measuring the dioxin in the workers' fat tissue. The least exposed workers, who were exposed for fewer than twenty years, had a relative risk for all cancers of 1.1 (confidence range 0.8 to 1.4). The highest exposed workers, who were exposed for over twenty years, had a relative risk of 2.6 (confidence range 1.2 to 4.9). Lung cancers were elevated in these workers. There were no soft tissue and connective tissue cancers reported, but because of the small size of the study, statistically, less than one case of such cancers was expected (Manz, 1991). A modest number of female workers was included in this study. The relative risk for all cancers was not elevated, but there was an increase in breast cancers. However, only 7 percent of the women worked in high-exposure departments.

The third important study was made on another group of German workers. This study examined 247 workers who were employed at a chemical manufacturing facility that produced 2,4,5-trichlorophenol contaminated with dioxin. This study found that workers with the highest exposure, as measured by an employment tenure of twenty years or more, had twice as much chance of getting cancer as people who didn't work in the plant. The relative risk was 2.0 with a confidence range of 1.2 to 3.2. The relative risk for lung cancer was elevated. This study also showed the highest cancer risks for the people with the highest exposures (Zober, 1990).

Other studies, although useful, all suffer from a lack of good exposure data. A large study of all cancers of over 18,000 European workers, including 1,500 women, was conducted by the International Agency for Research on Cancer (IARC). Despite problems in the definitions of exposure and in the length of followup, increases were seen in several forms of cancer, including lung cancer and soft tissue sarcoma. Some decreased breast cancer was found in females (Saracci, 1991). Several other studies have been conducted on worker groups (e.g., in Denmark, Britain, the Netherlands, and Sweden) showing isolated excess risk for multiple myeloma and cervical cancer in women and malignant melanoma in men. In general, the industrial studies point to increased risk for connective and soft tissue cancers, lung cancer, and for all cancers combined, especially for persons with relatively high exposures and long followup times. Followup time is the time between when exposure occurred and when the study looked for cancer. The longer the followup time, the greater the likelihood of observing cancer.

Three studies have been conducted on pulp and paper mill workers who have potentially been exposed to high levels of dioxin. No strong evidence was found for increased non-Hodgkin's lymphoma, lung cancer, or stomach cancer, but these studies were of short duration and had problems estimating exposures. Studies have also been conducted on pesticide applicators, railroad workers, and other workers who may have been exposed to dioxin or dioxin-like chemicals. Unfortunately, the exposure information is so poor that these studies contribute little or nothing to the knowledge of the toxic effects of dioxin.

Many studies have examined cancer incidence in Vietnam veterans due to their exposure to Agent Orange (a mixture of the herbicides 2,4,5-T and 2,4-D, both contaminated with dioxin). (See the Introduction, "The Politics of Dioxin.") Most of these studies were not able to determine whether exposure to Agent Orange caused cancer in Vietnam veterans. However, when a panel of scientists brought together by the National Academy of Sciences, Institute of Medicine, re-examined the cancer studies on Vietnam veterans, the panelists concluded that there was a positive association between exposure to herbicides and soft tissue sarcoma, non-Hodgkin's lymphoma, Hodgkin's disease, chloracne, porphyria cutanea tarda (a liver disorder). They also found suggestive evidence for respiratory cancers (lung, larynx, trachea), prostrate cancer, and multiple myeloma (NAS, 1993).

The EPA did report that measurements of dioxin levels in fat and blood were available for small groups of Vietnam veterans (Wolfe, 1990). To everyone's surprise, most of these men and women show levels of dioxin in serum and fat tissue indistinguishable from the general population. The exception is veterans belonging to a small, carefully defined military population that suffered high exposure. Even 1,261 veterans who participated in the actual spraying of Agent Orange in Operation Ranch Hand had dioxin levels only three times higher than controls (12.4 ppt compared to 4.2 ppt). There is a subgroup of Ranch Hand vets—"nonflying enlisted men"—whose dioxin levels are very high: 23.6 ppt. These are the men who did the spraying on the ground. These men are still quite young, so not much cancer has developed. Obviously, the studies on these veterans need to continue.

One study did identify a group of vets who went into areas that were being sprayed. These soldiers had an elevated risk of soft tissue sarcoma (Michalek, 1990). However, the number of cases reported was too small (in part, because the soldiers were so young) for any statistical significance.

Residents of Seveso, Italy, an industrial town north of Milan were exposed to high levels of dioxin as a result of an explosion in a chemical manufacturing plant. Many of them developed chloracne. The exposure was massive for some people, and thousands of residents were affected. The population has now been followed for at least ten years, which is not very long considering that most cancers take between twenty and thirty years to develop.

So far, increased cancer incidence has been found mostly for connective and soft tissue sarcoma among males, and for some relatively rare blood-related and liver cancers in both males and females. Previous studies have linked dioxin exposure to soft tissue sarcoma in humans. The cancers reported in this study may represent only the tip of the iceberg of the cancers that will result from this accident. Further followup is needed (Bertazzi, 1993).

In accidents in Yusho, Japan, and Yu-Cheng, Taiwan, healthy people ate food cooked with rice oil contaminated with PCBs and dibenzofurans. The exposures were massive, but lasted only a few months. The exposures in Japan occurred in 1968 and affected 1,900 people. In addition to the many non-cancerous effects, there was a statistically significant increase in liver cancers (9 observed compared to 1.6 expected) and lung cancer in males, with some erratic but statistically insignificant increases in females (Kuratsune, 1988).

In Yu-Cheng, Taiwan, in 1979, 2,000 people were exposed in a similar accident. There has not been enough time for cancer to develop among those who were exposed.

The studies described above looked at groups of highly exposed people and examined them for disease. Another approach to evaluating health problems in people is through "case-control studies." These studies are quite different in design from studies of highly exposed people. In case-control studies, one selects people with a disease (and suitable controls, who do not have the disease) and questions them about possible exposure to the chemicals being studied. If a higher proportion of the cases—in contrast to the controls—are found to have been exposed, the exposure is considered to be "associated" with the disease.

Case-control studies work "backwards" in that they try to identify a common exposure among people with a disease. If the people who have this disease also have a common exposure, or a higher exposure than those who don't have the disease, then a "link"

Seveso, Italy

On July 10, 1976, a runaway reaction during the production of trichlorophenol at the chemical plant ICMESA (Industrie Chimiche Meda Societa Azionaria), owned by Hoffmann-La Roche, resulted in the release of a toxic gas cloud containing dioxin into the surrounding community of Seveso, Italy. A whitish plume drifted into the immediate community of several thousand people. Altogether, some 37,000 people were exposed.

No one was at the plant at the time of the accident, so it was several hours before ICMESA even realized the accident had occurred. The closest residents were warned by a guard not to eat from their gardens, but no public announcements were made. ICMESA initially downplayed the accident, and only after scores of animals began to die five days after the accident did health authorities move in to investigate. Within three weeks, everyone closest to the plant was evacuated.

The most immediate effect of the accident was the death of domestic animals. Four percent died immediately.

The remaining animals (77,716) were slaughtered to protect the food chain.

The first sign of health problems in people was the development of skin lesions in children. The lesions developed into chloracne, a severe skin disorder commonly associated with dioxin exposure. Spontaneous abortions, immune system damage, and neurological disorders were reported but not found to be elevated (Homberger, 1979).

All of the studies evaluating health problems in Seveso residents have been criticized for the way exposure was determined. "Zones," or areas of contamination, were based solely on soil dioxin levels determined in tests done by ICMESA. For instance, some areas almost adjacent to the plant were considered to have low exposure because of low soil levels. Yet areas over a kilometer away were considered to have intermediate exposure because of the soil test results.

According to Harriet Rosenberg, sociology professor at the University of Toronto, who visited the Seveso site in April 1995, the mapping of the highest area of contamination "was and

is found. This contrasts with the ideal research situation, in which researchers look for more disease among exposed persons. However, when working with a rare disease—and most individual forms of cancer are rare—then both types of studies give just about the same answers.

The earliest studies on dioxin suggested that increases due to dioxin exposure might be found in soft tissue sarcoma and non-Hodgkin's lymphoma, and so case-control studies have largely concentrated on these two types of cancer. The best studies were conducted in Sweden and New Zealand on agricultural workers who were using the herbicide 2,4,5-T, which is contaminated with dioxin. These studies clearly showed that herbicide use was associated with an increase in soft tissue sarcoma. The relative risk varied from study to study, but the best studies showed a relative risk of 2.3 in Sweden (Eriksson, 1990) and 3.0 in New Zealand (Smith, 1983; 1984). Additional studies on agricultural populations from Nebraska, Kansas, Iowa, and Minnesota also showed increases in soft tissue sarcoma, but the risks were lower than in the Swedish studies. An excess risk of non-Hodgkin's lymphoma was observed in several studies.

Cancer in Hormone-Responsive Organs

Many of the toxic actions of dioxin can be explained by its interference with normal hormone functions. Hormones are potent

still is highly controversial in terms of the notion that it is possible to determine meaningful boundaries or that the risk declined from zone to zone. For example, there is a public housing project just outside the gates of the dumpsite [the plant has now been buried where it once stood] which was defined as being in Zone B. A resident of these buildings was quoted as saying at the time that it must have been a very smart cloud to make such a sharp turn just as it reached their homes" (Rosenberg, 1995).

chemicals. Very small amounts have profound biological effects on our development, reproduction, and well-being. Some environmental pollutants can act in the same way as hormones produced by the body. Although the clues have existed for several years, scientists have only recently recognized this as a problem. These substances, which include but are not limited to dioxin and dioxin-like chemicals, have been described as environmental estrogens, xenoestrogens, endocrine-disrupting chemicals, environmental drugs, or environmental hormones. Basically, these chemicals disrupt the actions of hormones of the endocrine system, particularly estrogen (the major female hormone), testosterone (the major male hormone), thyroid hormone, insulin, and corticosteroids. They may antagonize or counteract hormone action, or they may displace normal hormones from their receptor proteins.

Chemicals that act as environmental hormones are listed on Table 7-1. These substances are generally heavily chlorinated compounds with a ring structure, such as dioxin, DDT, and PCBs. Several of these chemicals, many of which are pesticides, have

---

Table 7-1

Chemicals That Act as Environmental Hormones

• 2,4-D	• Synthetic pyre-	• chlordecone (ke-
• 2,4,5-T	throids	pone)
• Alachlor	• Methoxychlor	• Aldicarb
• Atrazine	• Toxaphene	• DBCP
• Nitrofen	• DDT and metabo-	• PCBs
• Amitrole	lites	• Dioxins
• Metribuzin	• Carbaryl	• Pentachlorophe-
• Trifluralin	• Endosulfan	nol (PCP)
• Benomyl	• Mirex	• PBBs
• Mancozeb	• Transnonachlor	• Furans
• Zineb	• Chlordane	• Penta-to nonyl
• Metiram-complex	• Oxychlordane	phenols
• Maneb	• Dicofol	• Heptachlor and
• Ziram	• Dieldrin	heptachlor epox-
• Tributylin	• Parathion	ide
• Hexachlorobenz	• Methomyl	
ene	• Lindane	

Source: Colborn, 1993

---

already been banned (kepone, toxaphene, dieldrin, mirex, DBCP, DDT, PCB) because their toxic effects were so striking and so serious, even though the hormonal mechanism by which they exerted their activity was unknown. DDT was banned in 1972 partly due to eggshell thinning in birds, an effect that we now recognize as hormonal. Dibromochloropropane (DBCP) was banned because exposed workers became sterile (Whorton, 1983). Years after these chemicals were banned, scientists began to better understand their effects on the human body. Attention first focused on the issue of environmental hormones at a scientific conference on "Chemically Induced Alterations in Sexual and Functional Development" in 1992 (Colborn, 1992). This was followed by a BBC documentary called "Assault on the Male" in

1994, and an EPA conference in 1994 on "Estrogens in the Environment" (Birnbaum, 1994a).

What do chemicals that affect normal hormone function actually do? A chemical that interferes with insulin can alter glucose tolerance and increase the probability of diabetes, as dioxin does. A chemical that alters the sex hormones can reduce the ability to have children, cause birth defects of the reproductive system, or affect sexual behavior. Dioxin decreases the number of sperm and reduces the levels of testosterone. A chemical that acts as a hormone might also increase cancer risk in hormone-responsive organs. Breast cancer risk, for example, increases with increasing exposure to estrogen and chemicals that act like weak estrogens. This could explain some of the increases in hormone-related cancers that are occurring in industrialized society. For example, cancers in hormone-responsive organs such as the breast, testes, and prostate are increasing in the United States (Devesa, 1995; Santi, 1994).

For quite a while, environmentalists have predicted that the increasingly widespread use of chemicals in our society would cause an increase in cancer. And for quite a while, the cancer statistics proved them wrong. While the number of cancers in the country was increasing, this was explained almost entirely by the increasing number of older persons and by cigarette smoking.

Now, however, the predicted increase in cancer has arrived. Age-adjusted (meaning that the increased number of elderly people has been taken into account) cancer rates are increasing. The National Cancer Institute has compared the rate of cancers during a recent four-year period, 1987–1991, to the rate of cancers in a four-year period twelve years earlier, 1975–1979. All cancers increased 19 percent in men and 12 percent in women during this time (Devesa, 1995). Some of the increases in specific cancers for men are shown in Table 7-2, and for women in Table 7-3.

In men, prostate cancer increased 66 percent during this period. Part of the increase, but probably not all, is due to better diagnosis. Cancer of the testes has increased 34 percent in the same twelve-year period, and there has been no change in diagnostic methods. These two cancers in particular could be related to

Table 7-2
Age-Adjusted Rate of Cancers for Men, Per 100,000 Person-Years

Cancer	Rate in 1987-91	% increase since 1975-79
Total for all cancers	464	19
Prostate	121	66
Non-Hodgkin's lymphoma	19	60
Melanoma	15	67
Testes	5	34
Liver	4	33

Source: Devesa, 1995

exposure to chemicals, such as dioxin, that act as hormones. Men are also suffering increased cancer rates in hormone-responsive organs.

In women, two case-control studies indicate that pesticides can increase the risk of ovarian cancer, another hormonally mediated cancer. In Italy, one study showed an increased risk for ovarian cancer associated with exposure to herbicides (Donna, 1984), and another study showed a 2.7-fold increased risk for ovarian cancer with exposure to triazine herbicides for ten years or more (Donna, 1989). Cancer of the hormone-responsive organs has been associated with high fat in the diet. This is especially true for breast cancer, but vegetarians have long been known to have less cancer of the ovary and prostate as well. Since most of the chemicals that act like hormones reach humans through consumption of animal fat, it is possible that the recognized relationship between animal fat and some cancers could in fact be due to the presence of chemicals in the fat, and not to the fat itself.

The cancer incidence in children is increasing, and these increases are not explained by better diagnosis. In the same

Table 7-3

Age-Adjusted Rate of Cancers for Women, Per 100,000 Person-Years

Cancer	Rate in 1987-91	% increase since 1975-79
Total of all cancers	346	12
Breast	113	30
Lung	42	65
Melanoma	11	42
Non-Hodkin's lymphoma	12	35
Kidney	6	39
Liver	2	25

Source: Devesa, 1995

twelve-year period between 1979 and 1991, cancers have increased 13 percent in boys and 10 percent in girls, as shown on Table 7-4 (Devesa, 1995).

Breast Cancer

The increase in breast cancer is particularly significant because of the large number of women affected. During 1993, 182,000 new cases of breast cancer were diagnosed, and 46,000 women died of breast cancer. Over 2,600,000 American women currently have breast cancer. Fifty years ago a woman had a 1 in 20 chance of getting breast cancer; today that has increased to a 1 in 8 chance (Marshall, 1993). Breast cancer is the most prevalent cancer among women, surpassing lung and colon cancer. Some might argue that the increase in breast cancer is due to better diagnosis, through mammography, but breast cancer has been increasing at a steady 1 percent per year for a long time, ever since the advent of mammography. The fact that better diagnosis can not account for this increase is shown by another statistic. Some breast cancers are estrogen de-

Table 7-4

Age-adjusted Rate of Cancers for Children Ages 0-14, Per 100,000 Person-Years

Cancer	Rate in 1987-91	% Increase Since 1975-79
Total cancers-Boys	15.3	13
Total cancers-Girls	13.4	10
Brain-Boys	3.6	24
Brain-Girls	3.1	19
Leukemia-Boys	4.6	10
Leukemia-Girls	4.1	21
Bone-Boys	0.7	40
Bone-Girls	0.8	33

Source: Devesa, 1995

pendent (that is, they require estrogen to grow) and some are not. It is the estrogen-dependent breast cancers that account for most of the increase (Devesa, 1995).

The most important known risk factor for breast cancer is lifetime exposure to estrogen. Factors that increase exposure to estrogen increase the risk. Thus, an earlier age at puberty, a later age at menopause, a greater delay in having a first child, or not having a child can all increase estrogen exposure and breast cancer risk. But these changes alone are not enough to explain all the increase. Dr. Devra Lee Davis, with the Office of the Assistant Secretary of Health of the U.S. Department of Health and Human Services, has proposed that exposure to chemicals that act as weak estrogens will also increase breast cancer (Davis, 1993) and several studies support her suggestion.

One study that provides evidence linking breast cancer to chemicals that interfere with hormone functions was conducted by the New York State Department of Health (Melius, 1994). Long

Island, New York has one of the nation's highest breast cancer rates. And during the 1960s, Long Island was sprayed with 400,000 gallons of DDT, which still persists in the soil today. In a study of breast cancer in Long Island women, it was found that women who lived within one mile of a chemical factory had a 62 percent higher rate of breast cancer than women who did not live near a factory.

The evidence linking chemicals and breast cancer is also supported by a study that shows U.S. counties with a hazardous waste site are 6.5 times more likely to have an elevated rate of breast cancer than counties without a hazardous waste site (Griffith, 1989). In Israel, three carcinogenic chemicals that were heavily chlorinated were banned, and the rate of breast cancer decreased 30 percent (Westin, 1990). Women with breast cancer have 40 percent more DDE (a breakdown product of DDT that is also toxic and carcinogenic) in breast fat than women without cancer (Falck, 1992). In a separate study, women with breast cancer had higher levels of hexachlorocyclohexane (also known as lindane) in their blood than women without cancer (Mussalo-Rauhamaa, 1990). Yet another study showed that women exposed to PCBs have an elevated incidence of breast cancer (Kuratsune, 1987).

A very convincing study of the relationship between chemicals and breast cancer involved 14,000 women who were measured for the amount of DDE in blood and then followed for several years to determine who got breast cancer. Those women whose DDE levels were in the top 90 percent were four times as likely to get breast cancer as those women whose DDE levels were in the lowest 10 percent (Wolff, 1993). These studies in humans are supported by a number of animal studies showing that DDT, triazine, and atrazine increase mammary tumors in rats and mice (Scribner, 1981; Pinter, 1990).

The evidence from these studies is fairly strong that exposure to environmental hormones increases the risk of getting breast cancer. However, whether dioxin itself increases breast cancer is not clear. Dioxin is not considered to be an estrogen (Birnbaum, 1994a). The data on dioxin and breast cancer is quite confusing. At least one animal study (Kociba, 1978) and one human study (Bertazzi, 1993) report a decrease in breast cancer with dioxin exposure; yet another human study reports an increase in breast cancer (Manz, 1991).

What is clear is that dioxin suppresses the immune system, and any suppression of the immune system increases the risk of getting cancer. Perhaps this increased susceptibility, in combination with dioxin's other hormonal effects, could lead to an increase in breast cancer. The question of exactly what role dioxin plays in the development of breast cancer remains unanswered pending more scientific research.

Contributors

Dr. Beverly Paigen
Dr. Richard Clapp
Dr. Marvin Schneiderman
Stephen Lester, M.S.

Chapter Eight

Dioxin and Reproduction

D ioxin has far-reaching and serious effects on the reproductive system. It interferes with the activity of both male and female hormones. In addition, dioxin causes birth defects and endometriosis, a painful condition associated with menstruation. And, as shown in Chapter Seven, dioxin is one of a group of environmental toxins that interfere with hormones.

Endometriosis

E ndometriosis is a disease in which cells from the lining of the uterus grow in inappropriate places outside the uterus. This growth may occur on the ovaries, on the outside of the uterus, or in other places in the abdominal cavity. Why this occurs is not known. These new cells respond to the hormones of the monthly menstrual cycle just as the cells lining the uterus do: In the first part of the month, the tissue formed by the new cell growth thickens; at the end of the cycle, it bleeds.

Endometriosis causes pain during menstruation and during intercourse. It is also a common cause of infertility. The symptoms grow worse as the endometriosis spreads. More than five million women in the United States are affected by endometriosis. The number of women diagnosed with endometriosis appears to be increasing. This is partly due to better diagnosis and partly due to a real increase in the disease. From 1965 to 1994, there has been a 250 percent increase in hysterectomies for women with endometriosis between the ages of 15 and 24, and a 186 percent increase for women between the ages of 25 and 34 (Berger, 1995).

The discovery that dioxin can cause endometriosis was unexpected. From 1977 to 1982, different groups of monkeys were fed a diet with no dioxin, with 5 parts per trillion (ppt) dioxin, and with 25 ppt dioxin. The study ended and the monkeys were kept together with their colony. Then, from 1989–1992, three female monkeys who had participated in the experiment died due to blockage of the intestinal tract by severe endometriosis. Examination of the remaining monkeys showed moderate to severe endometriosis in 17 percent of the zero exposure group, 71 percent of the 5 ppt group, and 86 percent of the 25 ppt group (Rier, 1993). This finding has prompted a number of studies in humans to determine whether women with endometriosis have higher-than-normal levels of dioxins, PCBs, DDT, and other chemicals that affect hormone function. None of these additional studies has yet been completed or published.

Dioxin in Breast milk

Several studies have found high levels of dioxin in breast milk of nursing mothers (Schecter, 1994d). Dioxin is found in breast milk because breast milk is rich in fat and because dioxin tends to concentrate in fat tissue. In addition, animal studies have shown that, during breast-feeding, dioxin is transferred from the nursing mother to the infant, and that nursing infants can have a higher concentration of dioxin in their bodies than do their mothers. In one study of rhesus monkeys, from 17 to 44 percent of the dioxin in the mother was transferred to the infant during nursing (Bowman, 1989). In this study, dioxin concentrations in the fat of the infants were four times higher than in the fat of their mothers. In another study, of marmoset monkeys, breast-fed infants had higher concentrations of dioxin in their livers than did their mothers.

As discussed in Chapter Five, breast-fed babies have a higher intake of dioxin than any other population group. This is because breast milk is at the top of the human food chain, where dioxin concentrations are the greatest. In the EPA report, breast-fed infants are included in the group of "heavily exposed" populations. The EPA has issued assurances that the "heavily exposed" groups are small. Yet half of the American population has been breast-fed—not a small group at all. The EPA report provides no guidance for nursing mothers.

Nursing Infants

Nursing children is important. Breast milk is the perfect food for babies, the source of important immunities. Breast-fed babies are healthier than bottle-fed babies. They have fewer respiratory illnesses, fewer skin problems, cry less, have fewer allergies, and do not get constipated. Breast feeding is a special time of bonding for mother and child.

Although dioxin has been found in breast milk, we recommend that women continue to breast-feed their babies. There are clearly many benefits to nursing that outweigh most risks. What we need to do is to identify the sources of the dioxin that ends up in breast milk, and take the necessary steps to stop any additional exposures. We know that the banning of PCBs and DDT led to a reduction of those chemicals in human milk (Furst, 1994). Similarly, by stopping dioxin exposure, we will once again make breast milk safer.

Studies demonstrating the health benefits of nursing have been done in industrialized countries where dioxin would be expected to be in breast milk. So we do know that nursing babies is still a healthy thing to do. However, most studies on the health effects of nursing on babies are done during the period of infancy. What these studies do not reveal is some of the long term consequences of nursing—the more subtle changes that may not be detected for several years. For example, as described later in this chapter, sperm production in men has been declining over the last forty years. We know from animal studies that sperm production is reduced when pregnant rats are given dioxin (see next section). The male babies are exposed as fetuses and then exposed again through nursing. When they reach puberty, their sperm production is low. So some of the negative health consequences of nursing may be revealed only over the long term. This, again, is a question that could be studied scientifically by examining sperm counts of young men who were breast-fed and comparing them with the sperm counts of young men who were not breast-fed.

There may be women whose breast milk contains very high levels of dioxin and PCBs. These women may make a different choice about breast-feeding. This is a choice that a woman should never have to make. Yet, it is a choice that has been forced on women because of widespread environmental contamination by dioxin.

One alternative for a woman who is concerned and wants information is to find out the level of dioxin in her breast milk. The average level of dioxin in human breast milk in the United States is estimated to range from 16 to 20 part per trillion (ppt) of lipid (Schecter, 1992; USEPA, 1994d). But having a test done to determine the level of dioxin in breast milk is very difficult and expensive to do. There are fewer than twenty laboratories in the country that are competent to measure dioxin, and the test is very expensive, costing from $1,000 to $2,500 (Schecter, 1994a). One study showed that the concentration of dioxin was highly correlated to the concentration of PCBs (Furst, 1994), and it is much easier and cheaper to get breast milk tested for PCBs. For the time being, this may be a viable alternative for some women. The EPA and other federal agencies need to do more to reduce the costs of testing breast milk. The EPA should fund research to improve analytical techniques, which will reduce testing costs. In the meanwhile, the government could provide subsidies to women who cannot afford it to help pay the cost of having breast milk tested for dioxin.

A woman who is a long-term vegetarian or who has eaten a low-fat diet will have relatively low levels of dioxin in her breast milk. However, a woman who changes to a vegetarian diet at the time her baby is born is likely to have just as much dioxin in her milk as a woman who continues eating a typical American diet (Pluim, 1994). This is because the dioxin is coming from long-term fat "stores." Women have about the same concentration of dioxin in fat stores regardless of weight, but a woman who is overweight has a larger dioxin storeroom. Her breast milk may not have higher concentrations of dioxin, but it will take much longer for those concentrations to decrease. Women who are older and nursing a first baby will have higher levels of chemicals in milk simply because they have had a longer time to accumulate them (Bates, 1994).

Producing milk does accelerate the elimination of dioxin from a woman's body. A baby who has been nursed for a year is getting less dioxin in milk than he or she did at two months of age. After six weeks of nursing, the dioxin levels in breast milk are already measurably reduced (Rogan, 1987). In one woman exposed to these chemicals at her workplace, the level of PCBs in breast milk decreased from 14 parts per million (ppm) in her milk fat at the beginning of the nursing to about 4 ppm a year later (Sim, 1992). A second or third child who is breast-fed will also get less dioxin than a first child (Rogan, 1986). This reduction over time suggests one possible way of reducing the dioxin a baby gets. Between feedings, a nursing mother can pump her breasts and discard the milk. In this way, she will be speeding up the elimination of dioxin from her fat stores. The more milk she is able to pump and discard, the more dioxin she will be getting out of her body. If she changed to a low–animal fat or vegetarian diet at the same time, she would be reducing her intake of dioxin while increasing her elimination.

The government clearly needs to provide educational materials and support studies of nursing mothers, to help provide the facts women need to reduce dioxin exposure to their babies.

Effects of Dioxin on Male Reproduction

Table 8-1

What Dioxin Does to Male Animals

- Decreases testes size
- Decreases size of accessory sex organs: penis, prostate, seminal vesicle
- Lowers sperm count
- Lowers testosterone level
- Decreases sexual behavior
- Causes abnormal testes
  —degenerating sperm
  —loss of germ cells

Source: USEPA, 1994

Multiple animal studies show that dioxin has a profound detrimental effect on the male reproductive system. (See Table 8-1.)

Most of dioxin's effects on males result from deficiencies of the male hormone testosterone. Dioxin inhibits the synthesis of testosterone. Dioxin also affects the operation of the pituitary gland. Normally, when testosterone levels are low, the body has a feedback mechanism that sends a signal, using the luteinizing hormone (LH), from the pituitary gland to the testes to produce more testosterone. This mechanism depends on "imprinting" by testosterone surges during fetal and early life. Dioxin prevents this imprinting. Then, when the body produces more LH, the message that "more LH means produce more testosterone" is not recognized. No matter how high the LH goes, the body will not respond with more testosterone.

Pregnant rats were exposed to a single dose of TCDD in a series of studies (Mably, 1991, 1992, 1992a, 1992b). The rat pups were exposed in the womb and then again through the mother's milk. There were no signs of toxicity in either the baby rats or their mothers. However, changes in the males were observed at puberty. The sex organs were smaller, testosterone levels were lower, and their sperm counts were low. In addition, the descent of the testes was delayed (Peterson, 1993; Gray, 1993, 1995). When placed with a receptive female, these males took much longer to successfully mate with the female (they tried several times before achieving ejaculation), and once they had mated, there was a far longer interval before they were interested in mating again. Similar effects were observed in other studies.

When placed with another male, these dioxin-exposed males assumed a female mating position. The testosterone levels in these rats were only one-third of the levels in control rats, but were still high enough to cause masculine mating behavior (Demassa, 1977). It should be stressed that these changes in male behavior occurred long after exposure to dioxin had stopped. Therefore, the changes in spermatogenesis (sperm production), sexual behavior, and hormone levels were irreversible, much like birth defects.

The best explanation for these observations is that the rats failed to properly imprint male mating behavior during the testosterone surges that normally occur early in life. Sexual differentiation of the brain occurs during development, probably through changes in the hypothalamus. When dioxin is administered to rats, the highest concentrations are found in the hypothalamus and the pituitary. During early development, which includes late fetal life and early infancy, the presence of male hormones is required for imprinting the brain for male-type behavior. This includes imprinting male mating behavior, male-type response in the hypothalamus, and male-type biochemical response to hormones such as LH. Males and females differ in regulation of LH; dioxin-exposed males showed female-type regulation of LH. The male reproductive system is about 100 times more sensitive to dioxin when exposure occurs perinatally (just before and after birth) than when exposure occurs as an adult. These effects are not detectable until the offspring reaches puberty. According to the EPA report, these effects on male sexual behavior and on the male reproductive system are the most sensitive effects observed in rats exposed to dioxin.

Human studies support the observations seen in animals. In occupational or accidental exposures to dioxin, one common complaint among men is reduced sex drive and difficulty with erection and ejaculation. In the NIOSH study, dioxin levels in the blood were related to higher levels of LH and lower levels of testosterone (Egeland, 1994). This finding is consistent with results from animal studies, described above, that showed dioxin interferes with the LH signal that the body sends out when testosterone levels are low. In the study of Operation Ranch Hand veterans, those with the highest dioxin levels had the lowest testosterone levels, and had four times as many "unspecified testicular abnormalities" as did members of the control group (Roegner, 1991).

Decreasing Sperm Count

Several studies have shown that men today have about half the sperm count of their fathers (Carlsen, 1992; Irvine, 1994; Auger, 1995). The cause for this is not known, but several of these studies indicate that exposure to chemicals that act like hormones have interfered with normal hormone functions that control sperm production. Certain chemicals have been shown to lower sperm counts in animals and in exposed workers. In animal studies, dioxin (Mably, 1991; Nadler, 1969; Sedersten, 1978) and the pesticide endosulfan (Gupta, 1979; Singh, 1990) lowered testosterone levels and produced testicular atrophy in male rats. And chemicals such as dioxin (CDC, 1988b), kepone (Guzelian, 1982), 2,4-D (Lerda, 1991), and dibromochloropropane (Whorton, 1983) have been shown to reduce sperm count in men.

The decreased sperm count in men has been noted in several scientific reports, but up until 1992, none of these studies were convincing by themselves. Then, a group of Danish researchers used a statistical technique called "meta-analysis" to combine the results of all the studies. Meta-analysis is a statistical technique that allows one to combine many separate scientific studies to determine an overall effect. For example, suppose that one study of 50 men showed a 10 percent increase in sperm count, but another study of 1,500 men showed a 10 percent decrease. Instead of concluding that one study shows an increase and a second study contradicts the first, meta-analysis combines the studies to determine the average change in sperm count in all 1,550 men. The Danish researchers reviewed sixty-one published studies of nearly 15,000 European men and found that between 1938 and 1990, sperm count had decreased by 50 percent in one generation (Carlsen, 1992). Scientists hesitated to accept this conclusion, as

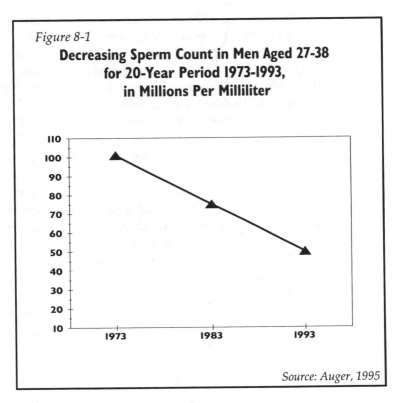

Figure 8-1

**Decreasing Sperm Count in Men Aged 27-38
for 20-Year Period 1973-1993,
in Millions Per Milliliter**

Source: Auger, 1995

the meta-analysis combined studies conducted in different countries at different times using different techniques.

Then, in 1995, a study was published that convinced many skeptics (Auger, 1995). One sperm bank in Paris that had been in existence for over twenty years examined the sperm count, sperm motility, and percentage of abnormal sperm in 1,351 healthy men. The sperm bank had detailed information on the age of each donor and the days of abstinence before donating sperm. The techniques for counting sperm had not changed over the twenty-year period. According to this study, sperm count, sperm motility, and the percent of normal sperm all decreased over the twenty-year period (1973 to 1992). The average sperm count declined by 2.1 percent per year from 89 million per milliliter (ml) in 1973 to 60 million per milliliter in 1992, a 33 percent decrease. Men who have less than 20 million sperm per milliliter are considered infertile.

The scientists further limited the analysis to young men between the ages of 27 and 38 who had abstained from sex for at least three days before donating sperm. Among this group of 382 men, the changes during the twenty-year period were even greater. Sperm count decreased by 2.6 percent per year, with a total reduction of slightly over 50 percent. Sperm motility decreased by 7 percent over the same time period, and the percentage of normal sperm decreased by 14 percent. The analysis was done by comparing the year of birth of each donor. This suggests that exposures during the fetal or early infant period, as shown in the animal studies, are most important in determining the number and quality of sperm produced at adulthood. Reduced sperm counts are associated with reduced fertility in men.

In this study, the authors found that the most critical factor in determining sperm count was the year of birth. They found that the quality of semen from fertile men of a given age in 1992 was significantly poorer than a fertile man of the same age in 1973. The sperm count in a thirty-year-old man born in 1945 was 102 million per milliliter, while the count in a thirty-year old born in 1962 was only 51 million (Auger, 1995). In other words, the year that the sample was taken was less important than the year the man was born.

A second important study, published in late 1994, examined the records of 3,729 semen donors born between 1940 and 1969 in Scotland. This study found a decline in sperm count from 128 million per milliliter in those born in the 1940s to 75 million in those born in the late 1960s—a 41 percent decrease (Irvine, 1994). Both these studies confirmed the earlier study showing that sperm counts have declined dramatically over the past fifty years.

Congenital Defects of the Male Reproductive Tract

Some birth defects involving the male reproductive tract have been increasing in industrialized countries (Giwercman, 1993). One is cryptorchidism, which means hidden testes. In nor-

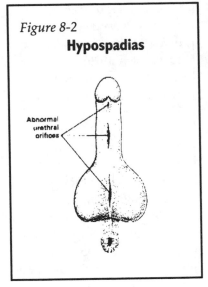

Figure 8-2
Hypospadias

Abnormal
urethral
orifices

mal development, the testes develop inside the abdominal cavity, just as the ovaries do. However, before birth the testes descend from the body cavity to rest outside the body below the penis. Sperm do not survive well at the higher temperature inside the abdominal cavity, so the testes are carried outside the body where the cooler temperature is more conducive to sperm production (Sharpe, 1993). An environmental toxin such as dioxin that disrupts normal male hormone function can cause failure of the testes to descend.

Another birth defect that is increasing is an abnormality in the male urethral opening called hypospadias (see Figure 8-2). This birth defect may also result from environmental chemicals that interfere with normal male hormone function (Giwercman, 1993).

Fetal Toxicity and Birth Defects

In Animal Studies

Dioxin is lethal to fish and birds in the early stages of their development. Young fish die from doses of dioxin that are 1/25th to 1/80th of the dose that kills adults (Walker, 1991; Kleeman, 1988). Similarly, bird eggs die from exposure to dioxin doses that are 1/100 to 1/200th of the amount it takes to kill an adult bird (Greig, 1973; Allred, 1977). Dioxin-exposed rats showed reduced fertility and a decrease in litter size (Murray, 1979). Dioxin also suppresses ovarian

The EPA's Long-Buried Evidence of Human Reproductive Hazards

In 1979, the U.S. EPA took emergency action to prevent human exposure to dioxin in the herbicide 2,4,5-T. An EPA study linked spontaneous abortions to forest spraying of 2,4,5-T by the federal government in a 1600-square-mile area of western Oregon.

The EPA study was the first large-scale human study to examine the reproductive effects in women of chronic, low-level dioxin exposure; it was also the first human study to correlate toxic effects directly to dioxin levels in water and food sources, human milk, wildlife, and human tissue.

One of the subjects of the study was Irene Durbin, who lived with her family along a tributary of the Alsea River. In early 1979 Irene was informed by the lead scientist for the study that her family's water supply was contaminated with 20 parts per trillion dioxin, that the water was unsafe, and that the scientist himself would not drink it. Irene's family had been drinking that water and eating local game and produce for seven years, during

function, suppresses the estrous cycle, and stops ovulation (Kociba, 1976; Barsotti, 1979; Allen, 1979).

In one study, female monkeys were fed 0, 5, and 25 parts per trillion (ppt) dioxin in their diet. There were no obvious signs of toxicity in these monkeys (although many years later they got endometriosis, as described earlier in this chapter). Of the group fed 5 ppt, 7 of 8 females conceived, with 6 resulting in live births, and one in a stillborn. This was not statistically different from the control group, where 8 of 8 gave birth to living babies. However, in the group fed 25 ppt, only 1 of the 8 females gave birth to a live baby. Scientists concluded that dioxin decreases fertility, makes it difficult to maintain pregnancy, and alters the ovarian cycle (Bowman, 1989).

It is clear that dioxin reduces the amount of the female hormone estrogen, which is produced primarily in the ovaries and acts to regulate certain female reproductive functions, but it is unclear exactly how dioxin does this. Dioxin does decrease estrogen receptor concentrations and may increase the metabolism of estrogen. Both of these actions would lower the effective estrogen concentration.

Animal studies have shown that dioxin also causes birth defects. Exposed chicks had malformations of the heart (Cheung, 1981; 1981a); exposed newborn rats had incomplete or absent vaginal openings (Gray, 1993); exposed newborn rats (Khera, 1973) and mice (Courtney, 1971; Neubert, 1972) had cleft palates and hydronephrosis. Hydronephrosis is a birth defect in which excessive growth on the lining of the ureter blocks the flow of urine, which then backs up and causes kidney damage.

In an important study on mice, a low dose of PCBs administered during fetal life caused hyperactivity, impaired learning, and caused difficulty in walking a wire, head bobbing, and circling (Tilson, 1979). Brain pathology showed decreased dopamine levels (which indicates altered brain function) and alterations in nerve conduc-

which time the entire surrounding watershed, as well as the adjacent roadside, were sprayed annually with dioxin-contaminated 2,4-D and 2,4,5-T.

During this period, members of the family suffered two miscarriages, a premature birth with numerous complications, chronic respiratory and digestive disorders, a fatal case of liver cancer, and the death by heart attack of Irene's husband at age thirty-eight.

The emergency ban of 2,4,5-T did not end the suffering of the Durbins and their equally devastated neighbors, who became unwitting guinea pigs in an unconscionable EPA experiment. The government substituted 2,4-D in subsequent defoliation spraying over their valley. In the wake of this spraying, every first trimester pregnancy in the valley spontaneously aborted, three children nearly died of spinal meningitis, one baby was born without a brain (anencephalic), a kitten was born with four eyes, and nearly every household reported outbreaks of rashes, uterine hemorrhaging, intestinal disorders, and respiratory disease.

Following this disaster, a preliminary study by the U.S. Centers for Disease Control found a 13-fold increase in neural tube birth defects in the Alsea study area, and post-2,4-D sampling identified dioxin in valley water supplies, wildlife, and tissues of the anencephalic baby. Sediment from the Durbin's water supply, resampled four years after 2,4-D replaced 2,4,5-T, contained 79 parts per trillion 2,3,7,8-TCDD, four times the 1979 level. Nevertheless, the EPA informed residents that this level presented no "immediate health hazard" and ended all further investigation.

The "smoking gun" dioxin samples of the Alsea study confirmed the link between dioxin exposure and the appalling human effects suffered by the Durbins and others in the study. But the Reagan EPA buried the most incriminating data of the Alsea study and ordered the scientists conducting the study not to publish their data nor discuss the study in any public forum.

While sabotaging its own study and muzzling its scientists, the Reagan EPA reached a secret agreement with Dow Chemical and other Agent tion. This study has not yet been repeated with dioxin.

Monkeys exposed to dioxin during fetal development and through breast milk had impaired cognitive function, and the dioxin-exposed infants clung to the mothers much more, an effect also seen in lead-exposed infants. This increased clinging has been interpreted to mean that the mothers had to provide more care to the infants (Schantz, 1986).

By examining all the animal studies in which dioxin was given prenatally, the following conclusions may be reached: (1) at the lowest doses, learning behavior and male reproduction are affected; (2) at somewhat higher doses, birth defects occur; and (3) at still higher doses, prenatal death can occur.

In Human Studies

Direct comparisons between animal studies and human studies related to fetal toxicity and birth defects are difficult because they have slightly different emphases. For example, the ani-

Orange manufacturers to allow 2,4,5-T back on the market, with "acceptable" levels of dioxin. This plan to remarket a known source of dioxin was scuttled by the scandal that erupted when the suppressed Alsea study dioxin analyses were leaked to the women subjects of the study. On October 14, 1983, in the midst of three separate investigations into the EPA's dioxin coverup, the EPA and Dow simultaneously announced the withdrawal and cancellation of all 2,4,5-T registrations. With the death of 2,4,5-T, the smoking gun evidence of dioxin's human effects was quietly buried in EPA archives.

Quietly buried as well were the EPA's steadily increasing data on the dioxin content of 2,4-D, thus allowing millions of pounds of 2,4-D products to be spread to this day on the lawns, parks, highways, schoolyards, playing fields, and farmlands of America.

—Carol Van Strum

mal studies have examined maternal exposure and its effect on offspring, while human studies have focused on primarily paternal exposure and offspring. The animal studies have examined spermatogenesis, fertility, sex organ development, and sexual behavior while the human studies have focused on birth defects and miscarriages.

When rice oil contaminated with PCBs and furans (both dioxin-like substances) was consumed accidentally in Yusho, Japan, and in Yu-Cheng, Taiwan, the effects on fetal development in humans were observed. In both cases, high mortality during infancy was observed, especially for babies born with hyperpigmented (over-colored) skin, another sign of dioxin toxicity. The mothers of almost all affected babies had chloracne, suggesting that the dose they were exposed to was quite high. In Yusho, exposed children also had reduced birth weights and retarded growth.

Tissues such as skin and nails are targets for dioxin. Babies in both cases had hyperpigmentation of skin and mucous membranes, abnormal fingernails and toenails, conjunctivitis, teeth in newborns, altered eruption of permanent teeth, and abnormally shaped tooth roots (Kuratsune, 1989; Chen, 1992). Accelerated tooth eruption was also seen in newborn mice exposed to TCDD (Madhukar, 1984). Similarly, several children born near the hazardous waste site at Love Canal had tooth abnormalities such as double sets of teeth and teeth in newborns (DOH, 1981).

The nervous system, like the skin, is derived from the ectoderm, the outermost of the three layers of cells found in an early embryo. Neurobehavioral effects often occur with prenatal exposure to dioxin (Mably 1991, 1992a). In Yusho and Yu-Cheng, exposed children had developmental delays, speech problems, behavioral difficulties, and impaired intellectual development (Rogan, 1988).

At Times Beach, Missouri, a study examined 402 births to mothers exposed to dioxin that was spread on roadways in contaminated waste oil used to control dust. Compared to unexposed mothers, the births to exposed mothers had increased fetal deaths, infant deaths, low birth weight babies, and birth defects (Stockbauer, 1988).

Vietnam Vets and Reproduction

Both the Vietnamese people and American soldiers were exposed to varying amounts of the herbicide Agent Orange as it was sprayed, and as it worked its way up the food chain. Unquestionably, some veterans were highly exposed to Agent Orange, but little data exists that identifies which soldiers were exposed. As described in the Introduction, this lack of information is a result of the botched efforts by different government agencies to carry out health studies. In spite of many problems with the various studies, it is clear that the spraying of Agent Orange did cause an increase in miscarriages and birth defects among those who were the most exposed.

In North Vietnam, the birth defect rate in children whose veteran fathers never served in South Vietnam was 6 per 1000 births; for those babies whose fathers did serve in South Vietnam, the rate was 29 per 1000 (Hatch, 1984). In another very large and stringent study of North Vietnamese veterans involving over 121,000 pregnancies, spontaneous abortions were increased significantly and birth defects were increased slightly, but the increase was not statistically significant (Hatch, 1984; Constable, 1985). A study of American Vietnam veterans carried out by the Centers for Disease Control found an elevation of spina bifida, cleft lip, and cleft palate, hydrocephalus, and childhood cancers (Erickson, 1984). The American Legion's study found an increase in miscarriages of fetuses fathered by Vietnam veterans (Stellman, 1988). All these studies were done on male veterans. Some nurses and other women serving in Vietnam have reported increases in miscarriages, but no scientific studies have been done on this group.

The blood dioxin levels were measured in several small groups of veterans. Veterans who handled Agent Orange regularly had higher levels of dioxin compared to other veterans. The median dioxin level for these vets was 25 picograms per gram (pg/g) of blood fat, compared with 3.9 pg/g for controls (Kahn, 1988). Veterans in Operation Ranch Hand, considered the most highly exposed group of soldiers, had 12.4 pg/g of dioxin, compared with controls with 4.2 pg/g. Among this group were some very highly exposed men who had levels as high as 166 pg/g of dioxin, compared with 10.4 pg/g for controls. Ranch Hand vets who became parents after they were in the service had children with more birth defects than those who had children before they went to Vietnam (Wolfe, 1992). Miscarriages were elevated in some groups but the increases did not match well with the blood levels of dioxin.

The studies of Vietnam veterans have many deficiencies. Even when the exposure data was available, as in the Ranch Hand study, dioxin levels were measured many years after exposure, and in most cases, years after the pregnancies occurred. No back calculation was done to estimate the likely dioxin levels when the pregnancies occurred. However, the elevated miscarriages and birth defects among children of the most exposed Vietnam veterans is consistent with animal studies and other studies of dioxin-exposed humans.

The Effects of Hormone-Disrupting Chemicals on Wildlife

Dioxin and many other chlorinated chemicals interfere with the normal function of the body's hormone system. Many of the health effects caused by these hormone-disrupting chemicals are associated with increased exposure to estrogens, as discussed in Chapter Seven, "Dioxin and Cancer."

Although dioxin is not estrogenic, it affects many endocrine systems. Dioxin can lower the levels of androgens (androgen is a generic name applied to male hormones; testosterone is the best known example of an androgen); affect the amount of thyroid hormones in the body; decrease insulin levels and change the amount of glucocorticoids; and affect digestive hormones and melatonin, the hormone that controls the daily rhythms of the body (Birnbaum, 1994a).

Dioxin's hormonal effects also influence development and reproduction. Some of the best evidence of these effects was first noted in wildlife studies. Several important studies are summarized here. As discussed earlier, dioxin is often not measured in environmental research studies because of the analytical difficulties and high cost of doing so. Even though the adverse health effects observed in wildlife have been linked to DDT and other chlorinated pesticides and chemicals, dioxin is also likely to be present in these animals and contributing to observed health effects as well. DDT and dioxin both interfere with hormone functions which result in adverse health effects.

Lake Apopka is Florida's third largest freshwater lake. A Superfund site is located along the shore of the lake, and a large spill of the pesticides DDT and dicofol from that site has contaminated the lake. The alligator population in the lake has declined by 90 percent since the 1970s. Scientists studied alligator eggs from

Lake Apopka and a nearby uncontaminated lake. Not surprisingly, eggs from Lake Apopka contained high concentrations of DDE, a DDT breakdown product. Eggs were incubated in the laboratory, allowed to hatch, and the baby alligators observed for six months. Seventy-two percent of Lake Apopka eggs never hatched because the embryos died, compared with 48 percent of control eggs from an uncontaminated lake. Death among young alligators was 41 percent for those from Lake Apopka compared with 1 percent for controls.

Twice as many females as males were born from Lake Apopka eggs. Young female alligators had elevated estrogen levels and males had decreased testosterone levels compared with controls. More importantly, sexual development was abnormal, so future reproductive success is seriously compromised. Only sixteen of twenty-five alligators from Lake Apopka eggs survived to the end of the six-month experiment (Guillette, 1994).

Microscopic examination of the testes and ovaries revealed that the scientists had misdiagnosed the sex of 25 percent of the alligators based on the external genitalia. Some of the males had no penis; some of the females had a penis-like structure. The testes were poorly developed, and the ovaries were abnormal (Guillette, 1994).

At first, scientists concluded that the pollutants in the lake had an estrogenic effect—that is, they acted like female hormones. But a recent study by the EPA, presented at a 1995 American Society of Zoology meeting, indicated that DDE exerts its feminizing effect by blocking male hormones (Culotta, 1995). This study showed that DDE binds to both rat and human androgen receptors and thus blocks the action of androgen. The alligators were not the only wildlife in trouble in Lake Apopka. Most of the turtles caught were female or intersex (their sex organs were partly female and partly male). The environmental toxins had a feminizing effect on the turtles, as they had on the alligators (Bergeron, 1994). The sex of turtles, alligators, and other reptiles is determined quite differently than it is for mammals. The temperature at which the eggs develop has a strong effect on the sex of the hatchling. Higher temperatures result in more females being born. Exposure to environmental chemicals such as DDT, dioxin, and other dioxin-like chemicals has an effect similar to that of a higher temperature, which results in more females being born. Laboratory studies of PCBs applied to turtle eggs have also shown that the application of certain PCBs resulted in almost all the eggs developing into female turtles.

The decline in bald eagle, peregrine falcon, and other bird populations at the top of the food chain due to eggshell thinning is something that has been known about for a long time. Now that DDT has been banned, bald eagles are making a comeback in most parts of the United States, and the bird will probably be removed from the endangered species list in the northern part of the United States. However, bald eagles along the coast of Maine and in other shoreline areas such as the Great Lakes (Colborn, 1990) and the Columbia River (Anthony, 1993), are not reproducing well. In Maine, eagle eggs that do not hatch contain high levels of dioxin and PCBs. Nesting pairs of bald eagles that arrive at the shores of the Great Lakes from inland are able to nest and have chicks for the first two years. However, the consumption of contaminated fish from the Great Lakes eventually leads to reproductive failure, and their dead eggs contain high concentrations of DDT (Wiemeyer, 1984). These dead eggs also contain PCBs, chlordane, dieldrin, and other chemicals that act like hormones. Dioxin was not measured in this study.

Because these chemicals are persistent and accumulate in fat, they are at especially high concentrations in egg yolk and breast milk, thus affecting the developing embryo and newborn at a time when the toxic effects are more detrimental. These chemicals lead to reduced fertility, lowered hatch rate of eggs in fish and birds, and lowered viability of offspring, posing a significant threat to wildlife (Colborn, 1993).

Contributors

Dr. Beverly Paigen
Stephen Lester, M.S.
Carol Van Strum

Chapter Nine

Other Health Effects

Dioxins, especially TCDD, are toxic to many species of animals. The symptoms of short-term, high-level exposure ("acute toxicity") in laboratory animals include liver damage, shrinking of the thymus, loss of body weight, changes in the skin, reduced prostate weight, reduced uterine weight, reduced testicular size, porphyria (overproduction of a protein related to red blood cell formation), bleeding in various organs, increased thyroid weight, damage to the adrenal glands, changes in heart muscle, and changes in the amount of protein and fat in the blood.

When small doses of dioxin are given to animals in their food over several months or years, a number of additional health problems are observed. The symptoms of this "chronic toxicity" include:

- increased levels of certain proteins known as cytochrome p450 and mixed-function oxidases;

- altered enzyme levels in the liver, indicative of liver damage;

- changes in skin similar to the chloracne seen in humans;

- reduced numbers of estrogen receptors in the liver;

- reduced testosterone levels;

- decreased ability to store Vitamin A;

- changes in lipid function; and

- suppression of the immune system.

The health effects of dioxin on humans have been more difficult to determine. Several early studies on dioxin-exposed people were flawed in many ways. (See the Introduction and Chapter Seven.) The most reliable information on human health effects comes from three excellent studies that measured serum (blood) or fat levels of dioxin to determine exposure: the NIOSH study of U.S. chemical workers (Sweeney, 1989), the Ranch Hand study of Vietnam veterans (Roegner, 1991), and studies of the residents of Times Beach, Missouri (Webb, 1989). These studies revealed a wide range of toxic health effects of dioxin exposure in addition to cancer and reproductive problems.

Chloracne and Other Skin Effects

Chloracne is considered the hallmark effect of exposure to dioxin, showing up more often than any other symptom following exposure. It is a serious skin disorder that begins with acne-like skin eruptions within one to two months of exposure. The disease progresses to more severe symptoms, characterized by pus-filled boils, pimples that may be colored more darkly than the rest of the face, blackheads, and cysts that sometimes persist for years (Reggiani, 1980). In some individuals, the condition disappears after exposure ceases, but in other cases, it remains for more than ten years (Suskind, 1984). One study reported an average duration of twenty-six years (Moses, 1984). Dioxin can also cause hirsutism, an abnormal distribution of facial hair; and hyperpigmentation, the overcoloration of the skin.

The Gastrointestinal System

Increased liver size is consistently reported in animals exposed to dioxin. Increased liver size has also been reported in dioxin-exposed people, primarily in case reports of accidental exposure, and it usually subsides over time. The liver can be considered the "powerhouse" of the body. It is where all foreign substances such as chemicals are processed or "metabolized" in an effort to rid the body of these substances. An enlarged liver is a measure of a liver that is overworked, indicating that the body is overloaded with toxic substances.

Gamma glutamyl transferase, or GGT, is a liver enzyme that can spill into the blood when the liver is damaged. It is often included in standard clinical biochemistry screens. Increases in GGT were found in animal studies and in some human dioxin exposures, particularly in Seveso, Italy (Mocarelli, 1986), and in a study of trichlorophenol production workers (Calvert, 1992). These same workers had an increased prevalence of ulcers, as did dioxin-exposed workers at Dow Chemical in Midland, Michigan. However, Ranch Hand veterans and the NIOSH study of workers did not have more ulcers with increasing dioxin levels.

Thyroid function

Several of the toxic effects of dioxin and dioxin-like PCBs resemble hypothyroidism, a reduced functioning of the thyroid gland. PCBs and dioxin are structurally similar to the thyroid hormone. Several animals studies confirm that exposure to PCBs or dioxin causes a reduction of thyroid hormone (Stone, 1995a). Very few human studies have examined the effect of these chemicals on thyroid function, and most of these studies, which were of adults, showed no change. However, one study of breast-fed babies showed reduced thyroid function in those whose mothers had the highest dioxin levels in their breast milk (Pluim, 1992).

Thyroid hormone is very important to human brain development. Deficiencies of thyroid hormone during fetal life or during early infancy can lead to mental retardation, hearing loss, and speech problems. Even if their IQ is in the normal range, children with thyroid deficiencies have language comprehension problems, impaired learning and memory, and hyperactive behavior. These problems are very similar to those experienced by the children exposed to PCBs and dioxins from contaminated rice oil used for cooking in the Yusho and Yu-Cheng incidents (Porterfield, 1994).

Diabetes

Dioxin interferes with the hormone insulin, alters glucose tolerance, and can lead to diabetes. In one study of fifty-five exposed workers evaluated ten years after exposure, 50 percent of the workers were diabetic or had abnormal glucose tolerance tests, an early sign of diabetes (Pazderova-Vejlupkova, 1981). Due to this remarkable finding, several other studies examined either abnormal glucose levels or diagnosed cases of diabetes in dioxin-exposed populations. Although the results vary from study to study, the most reliable studies consistently show that dioxin increases the risk of diabetes. In the NIOSH study, for example, the risk of diabetes increased 12 percent for every 100 picogram dioxin/gram (pg/g) of lipid in the blood (Sweeney, 1992). In the Ranch Hand study, veterans with blood dioxin greater than 33.3 pg/g had a relative risk of 2.5 for diabetes (Roegner, 1991).

In the EPA report, the agency expressed considerable concern that no animal studies showed increases in plasma glucose with dioxin exposure. Such a lack of correlation between animal studies and humans is a cause for concern. However, there is a rather simple explanation for the difference. Animal diets are low in fat—usually between 8 percent and 15 percent fat—and this fat is unsaturated: human diets, in contrast, contain between 30 percent and 50 percent fat, with a high proportion of saturated fat. If animals were fed a diet comparable to humans, and then tested for the effects on plasma glucose and diabetes, the discrepancies between human and animal studies might disappear. Dioxin simply has not yet been tested under the same dietary conditions for both animals and humans.

Table 9-1

Neurological Symptoms Reported After Dioxin Exposure

- headaches
- dizziness
- irritability
- depression
- insomnia
- tremors
- numbness
- anorexia
- muscle weakness
- nervousness

- anxiety
- crying spells
- apathy, fatigue
- slowed thinking
- social withdrawal
- trouble concentrating
- reduced sex drive
- difficulties with erection

- decreased mental efficiency
- depression
- loss of feeling in extremities
- tingling in toes and fingers
- slowed nerve conduction velocity

Source: USEPA, 1994a

The Nervous System

Dioxin causes neurological damage in animal studies. Many case reports or studies of dioxin-exposed people also show damage to the central nervous system or the peripheral nervous system, as shown in Table 9-1.

The EPA concluded, after reviewing all the neurological reports, that high exposure to dioxin causes neurological damage, but that these effects are short-lived. According to the EPA, studies conducted long after exposure ended, such as those on the Ranch Hand veterans, Seveso residents, or exposed chemical workers, indicate that neurological damage does not persist.

But the data in these reports is not consistent with the EPA's statement. In our reading of the studies, neurological damage persisted in some adults for ten years, especially in the chemical workers (Pazderova-Vejlupkova, 1981). The toxicity of dioxin to the developing nervous system might be more severe and could be permanent. The pregnant women who ate contaminated rice oil in Yu-Cheng had babies that were developmentally delayed and had speech problems. At ages eight to fourteen, these children still scored lower in IQ tests, had more behavior problems, and tended to be hyperactive (Guo, 1992, 1993).

The Circulatory System

A nimal studies clearly indicate that dioxin damages the heart muscle and affects the ability of the heart to contract. These changes lead to arrhythmias (irregular heartbeat or unexplained racing of the heart). In addition, dioxin damages heart valves in rats and alters cardiac function in rats and guinea pigs.

The human studies are not so clear, perhaps because of the "healthy worker" effect. The healthy worker effect is a special bias observed in occupational health studies, which results from the fact that an employed population is generally healthier than an unemployed or general population of the same age. Workers generally suffer less illness, including heart disease, and have lower death rates than the general population.

In dioxin-exposed workers, this "healthy worker" effect is missing. These workers suffer from heart disease at rates comparable to those of the general population. This suggests that they have an elevated rate of heart disease. To determine whether this is true, dioxin-exposed workers must be compared to other workers to account for the healthy worker effect.

Another problem is that studies of dioxin-exposed workers examine "survivors"—that is, the people who have survived the illnesses caused by exposures. Some of the sickest people (and most highly exposed) have died, or have severe illnesses, and are no longer in the workforce. In studies that only examine survivors, the risk of diseases that cause death or severe disability are underestimated. In spite of these problems, some human studies show that dioxin exposure does cause heart disease.

● Ranch Hand Veterans with the highest dioxin levels had higher blood pressure, more arrhythmias, and more abnormal pulses (Roegner, 1991).

- Ranch Hand veterans and Australian Vietnam veterans had a slightly elevated risk of heart disease compared with non-Vietnam veterans (Michalek, 1990; Fett, 1987).

- In Seveso, heart disease was increased in males and females, although the increase was statistically significant only in males (Bertazzi, 1989).

- Cerebrovascular disease (stroke) was elevated among trichlorophenol workers in Holland (Buenode Mesquita, 1993) and Michigan (Bond, 1987) and among males in Seveso (Bertazzi, 1989).

- Dioxin causes an elevation in plasma cholesterol and triglycerides, which increase the risk of heart disease (Martin, 1984; Moses, 1984; Roegner, 1991).

The EPA report stated that the information on whether dioxin increases cholesterol is confusing. This was because veterans in the Ranch Hand study showed elevations in cholesterol levels long after exposure stopped, while the residents of Seveso, Italy, showed no elevation shortly after the incident occurred. There is an obvious explanation for these conflicting results. The human studies showing that dioxin increases cholesterol were conducted in countries with diets high in saturated fats (the United States and the former West Germany); the study showing no increase was conducted in a country where unsaturated olive oil, rather than butter, is the fat of choice (Italy). Apparently, dioxin causes blood cholesterol to rise if the diet is already high in saturated fat.

The Lung

Long-term exposure to dioxin in rats, mice, and monkeys resulted in changes in the lung that are consistent with chronic bronchitis in humans. In human studies, acute exposure to dioxin caused respiratory problems (Zack, 1980; Goldman, 1972). The effect of long-term exposure on the lung has been observed only

Table 9-2

Overall Health Effects Associated with Dioxin and Dioxin-like Chemicals

Cancer:

- Soft or connective tissue, lung, liver, stomach, Non-Hodgkin's lymphoma

Male reproductive toxicity:

- Reduced sperm count
- Testicular atrophy
- Abnormal testis structure
- Decreased testis size
- Decreased sex drive
- Alterations in male hormone levels—decreased testosterone, androgen; increased follicle-stimulating hormone (FSH) and leutinizing hormone (LH)
- Feminization of hormonal and behavioral responses

Female reproductive toxicity:

- Hormonal changes
- Decreased fertility
- Adverse pregnancy outcomes—miscarriages; inability to maintain pregnancy
- Ovarian dysfunction—suppression of the estrous cycle, anovulation of the menstrual cycle
- Endometriosis

Effects on unborn fetus:

- Birth defects—cleft palate, hydronephrosis

- Alterations in reproductive system
- Decreased sperm count
- Altered mating behavior
- Structural abnormalities in female genitalia
- Reduced fertility
- Delayed puberty
- Neurological problems
- Developmental problems

Skin disorders:

- Chloracne
- Hyperpigmentation
- Eye lid cyst
- Hypertrichosis
- Actinic keratosis
- Peyronie's disease
- Hirsutism

Metabolic and hormonal changes including:

- Altered glucose tolerance and decreased insulin levels leading to increased risk of diabetes
- Altered fat metabolism leading to elevated lipids, cholesterol and triglycerides, and increased risk of heart disease
- Altered porphyrin metabolism leading to porphyria cutanea tarda (PCT) and elevated urinary porphyrins including uropor-

phyrin, urobilinogen, co-proporphyrin
- Weight loss, wasting syndrome
- Changes in thyroid hormones

Damage to central and peripheral nervous systems leading to:

- Increased irritability and nervousness
- Decreased pin prick sensation
- Impaired neurological development and subsequent cognitive deficits

Damage to liver as measured by

- Elevated Gamma Glutamyl Transferase (GGT) levels
- Enlarged liver
- Elevation of liver enzymes other than GGT (LDH, AST, ALT and D-glucaric acid
- Cirrhosis

Damage to the immune system leading to:

- Reduced size of thymus
- Increased T-4 cells
- Increased ratio T-4 to TBG cells
- Increase susceptibility to infectious disease
- Increased risk of cancer

Lung problems:

- Irritation
- Tracheobronchitis
- Decreased lung function

Other gastrointestinal damage leading to:

- Loss of appetite
- Nausea
- Damage to heart leading to circulatory disorders and heart disease

Source: USEPA, 1994a, 1994b; DeVito, 1994

in the study on Ranch Hand veterans. Those veterans with the highest serum dioxin levels had reduced lung function (Roegner, 1991). These changes were small, but statistically significant.

Conclusions

The EPA examined the evidence of each of the health effects of dioxin, taking into account the number of scientific studies, the quality of the studies, the sequence of events (exposure and appearance of illness), the degree of exposure, the consistency among studies, and biologic plausibility (whether the disease observed fits with the known biochemical actions of dioxin). From all this information, the EPA determined which effects are *known* to be caused by dioxin, which effects are *probably* caused by dioxin, and which effects are *likely* to be caused by dioxin. An overall summary of many of the health effects associated with dioxin and dioxin-like chemicals is shown in Table 9-2.

Contributors

Dr. Beverly Paigen
Dr. Richard Clapp
Stephen Lester, M.S.

Part Two

Organizing

Chapter Ten

Organizing to Stop the Poisoning

The truth won't stop the poisoning. But organizing will.

In Part One, we've presented all the facts we've found about dioxin. We've documented how dioxin from incinerators and cement kilns is poisoning our food and our breast milk. We've described in detail how dioxin interferes with our bodies' most basic functions, leading to diabetes, endometriosis, immune system suppression, infertility, and cancer. We've explained how the American people are full or nearly full of dioxin, making it a public health necessity that we stop dioxin pollution at its sources—immediately. But all of the evidence in all of the studies about the tremendous harm that dioxin is doing to our health won't, by itself, stop the poisoning.

The truth is only a start. In order for things to change, the truth has to be understood by a large group of people who then use this knowledge to fuel their efforts to win justice. The truth won't stop the poisoning, but organizing will. That's why Part Two of this book is about organizing. It outlines how you can help stop the poisoning by organizing a Stop Dioxin Exposure Campaign in your own community—and it tells you what to expect when you do.

According to Webster's dictionary, *organizing* means "uniting in a body or becoming systematically arranged." Organizing to stop dioxin exposure means uniting, and arranging, and more.

In 1963, Martin Luther King, Jr. spoke to a mass meeting in Birmingham, Alabama. He spoke of the 2,500 people who were in jail in Birmingham for protesting segregation:

> They are carving a tunnel of hope through the great mountain of despair. They will bring to this nation a newness and a genuine quality and an idealism that it so desperately needs.

Organizing is how we carve the tunnel of hope through the mountain of dioxin despair, and stop the poisoning.

Organizing is how we restore the balance between the rights of the people to safe food and healthy communities, and the rights of corporations to profit and pollute. We will never have as much money as the corporate polluters. We will never be able to afford their Madison Avenue media campaigns, or their twenty-four-hour access to elected officials. But we can build our own power to overcome the influence of corporate power. We can do this by organizing to demonstrate the strength of our numbers and the righteousness of our demands.

Organizing is about speaking truth to power. Successful organizing happens when a group of people finds visible ways to use the truth to wake up the conscience of a larger group. In an era when politics is defined by scandals and sound bytes, organizing can remind the American people that political life is supposed to be about self-government, justice, and the common good.

Organizing to stop dioxin exposure is fundamentally organizing to rebuild democracy. We can't change the corporate decisions that result in dioxin exposure without challenging the dominance corporations now have over public life. Our campaigns must not be only about the danger of dioxin, but also about the dangers of a society where money buys power. To create the equality and justice of a true democracy, our organizing must restore the people's inalienable right to govern and protect themselves.

The United States was founded on the principle of self-governance. Revolutionary-era patriots opposed the rule of a distant monarch and parliament. Instead, they designed a new system of government in which the people represent and rule themselves.

Two hundred and twenty years later, the American people feel less and less control over government. Going into a voting booth may still evoke the pride of fulfilling a civic duty, but how often have you felt that you are helping to elect true representatives of the people? Most of the time voting seems like picking the lesser of two or more evils. Even when we identify people we're proud to elect, they too often end up disappointing us once in office. Money takes over, and the corporations show us who is really in control.

The American people never agreed to give corporations control of our air, our food, our land, and our water. We didn't revolt against King George so we could have King Dow or Emperor WMX. In the campaign against dioxin exposure, we have the opportunity to engage the American people in a conversation about regaining our right to safe food; clean air, water, and land; and a country in which people, not corporations, control the citizens' lives.

The alienation of the American people is damaging every aspect of public life. Schools can't educate, courts

Jacksonville, Arkansas

One week before his election to the presidency of the United States, Governor Bill Clinton of Arkansas gave the final order to start burning dioxin in an incinerator in a residential area of Jacksonville, Arkansas, a community of 29,000 people fifteen miles northeast of Little Rock. The incinerator was built with $10.7 million in state funds for the purpose of burning 30,000 barrels of mixed hazardous wastes abandoned in Jacksonville by a company called Vertac Chemical, which manufactured pesticides and herbicides there from 1948 to 1986.

Over the years, the Vertac site was used for manufacturing DDT, aldrin, dieldrin, toxaphene, and the chemical warfare defoliants 2,4-D, Silvex, 2,4,5-T, and Agent Orange. In 1979 state and federal investigators discovered dioxin on the Vertac property and in soil and water several hundred yards from the site. Official surveys subsequently found dioxin from the plant in Jacksonville's central city park, making its once-popular swimming and fishing lake off-limits to the public.

In 1986, Vertac declared bankruptcy and willed its ninety-three-acre site to the people of Arkansas. Vertac's executives abruptly left town and have never been successfully traced. The *New York Times* said, "Vertac abandoned the plant leaving behind roughly 30,000 barrels of chemical wastes, along with acres of contaminated soil, tanks filled with toxic materials, and miles of poisonous piping. The EPA considers the site one of the country's worst hazardous waste sites, not only because of [the] extent of the contamination but also because the plant is only a few blocks from a day care center, a hospital, and hundreds of houses."

By 1989 state and federal officials had made firm plans to build the incinerator to burn some 22 million pounds of Vertac's dioxin-laced wastes. On at least two occasions a majority of the citizens of Jacksonville expressed, through referendums and public meetings, the fact that they did not want the incinerator built. Many local people considered it a dirty, dangerous way to "get rid of" the wastes. They pointed out that cleanup teams had already packed the 30,000 bar-

and police can't control crime, and demagogues use fear and prejudice to promote misguided "quick fixes" for complicated social problems. When people organize themselves to protect their communities from further dioxin contamination, they're also fighting back against the alienation and isolation that destroys the promise of democracy.

The Government Won't Stop the Poisoning, But Organizing Will

In a true representative democracy, we would expect that an agency set up to protect the environment would do just that. But today, the United States Environmental Protection Agency protects the right to pollute. This is another reason we must organize to protect ourselves against dioxin exposure—because no one else will do it for us.

The EPA's recommended water quality standard to protect aquatic life and preserve streams and lakes is .013 parts of dioxin per quadrillion (quadrillion equals

1,000,000,000,000,000). For safe drinking water, the recommended standard is 30 parts per quadrillion.

How can the EPA justify setting a standard for lakes and streams that is 2,000 times stricter than the standard for people? The answer is that setting standards is not about good science and the facts. It's about the corporations' pressure on government agencies to protect their financial interests. It's about "public" comment periods that are dominated by corporate scientists and consultants, public relations firms and lobbyists; it's about an agency that is pressured to justify every step with a level of scientific proof or certainty that has no place in a regulatory arena.

The proposed standards for medical and municipal waste incinerators are another example of the EPA's primary role as a permitter of—not a protector against—poisoning. Although the EPA recognizes that these types of incinerators are two of the greatest sources of dioxin in the country, the EPA's proposed standards "to protect public health" simply *limit* emissions by requiring air pol-

rels in special drums, which were not an immediate threat. The real threat was the wastes already released into the community, the ground, and the groundwater. State and federal officials ignored these expressions of sentiment, turned a blind eye to alternative technologies, and forged ahead with their plan to burn the visible evidence— to make Jacksonville look clean again. The real cleanup of soil and water would have to wait for a later time.

At public meetings throughout 1989 and 1990, state and federal environmental officials insisted repeatedly that the Vertac site incinerator would emit zero dioxin into the surrounding community.

A trial burn was conducted during October 1991. State and federal officials examined the data and declared the incinerator a resounding success. But Greenpeace chemist Pat Costner analyzed the trial burn data and published her own analysis, showing that the incinerator had not achieved the required 99.9999 percent destruction of the wastes, but had in fact achieved only 99.96 percent destruction. This meant that the incinerator was releasing

400 times as much dioxin as the regulations intended.

State and federal officials studied Costner's analysis and subsequently admitted that she was right. Further calculations showed that the Vertac incinerator would emit somewhere between 8 million and 45 million "EPA unsafe" doses of dioxin into the community during the two-year burn. State and federal officials said that the proposed burn would not violate any state or federal laws, and thus should be allowed to proceed. One Arkansas health department official excused the releasing of dioxin into the community, saying, "You have to appreciate how much dioxin there is in this community already." Pat Costner points out that this will be the largest intentional release of dioxin that has ever been executed.

The Vertac incinerator was shut down in February 1993 by U.S. District Judge Stephen Reasoner. EPA scientists had known since 1985 that incinerators cannot achieve 99.9999 percent efficiency in destroying wastes present in low concentrations, but EPA officials had stated at hundreds of public presentations since 1985 that lution controls. These standards don't mean incinerators will comply. And even if they all did, there still would be too much pollution.

The problem of dioxin is so serious that it cannot be addressed at the discharge pipe. With dioxin causing a wide range of health problems at such low levels, and with a general population body burden at or just below these levels, we can no longer afford to attempt to "control" pollution after it is generated. We must prevent dioxin pollution at the source. By organizing in Stop Dioxin Exposure Coalitions, we can provide our communities with the protections that EPA does not give.

Not only is the EPA a poor protector against dioxin, it actively participates in dioxin production. According to the EPA report, of the 50 percent of dioxin sources that are known, 53 percent are medical waste incinerators, 31 percent are municipal waste incinerators, and 4 percent are cement kilns. Yet the EPA continues to use incineration to clean up contaminated sites on the federal

Superfund list. In contrast to the 171 operating garbage incinerators and 191 hazardous waste incinerators in operation or in the permit phase, the EPA itself is operating or proposing to operate 276 incinerators across the country. The EPA's own studies show that incinerators cannot destroy and remove 99.9999 percent of dioxin (often referred to as six 9s), as required by EPA standards.

In many communities, the EPA has fought local residents' efforts to identify cleanup alternatives that don't create new public health threats. In New Bedford Harbor in Massachusetts, the EPA fought the group Hands Across the River, which opposed the EPA's plan to burn PCBs in a mobile incinerator. In Slidell, Louisiana, members of Slidell Working Against Major Pollution (SWAMP) is fighting the EPA's plan to keep burning chemical waste from Superfund sites in the town. After forty years of PCB and heavy metal contamination at the Crab Orchard Wildlife Preserve in southern Illinois, the EPA is determined to incinerate in "utter disregard for our concerns," according to Rose Row-

99.9999 percent could be achieved. Lying to the public carries no penalty, but lying to a judge is a different matter. At a hearing in Little Rock on February 12, the EPA was represented by U.S. Justice Department lawyer Ron Spritzer. Judge Reasoner said to Spritzer, "Indulge me for a moment. If I asked you to prove that you could achieve a six 9 DRE [99.9999 percent destruction and removal efficiency] on dioxin, could you physically produce technological data that shows that?"

"No sir, we could not," said Mr. Spritzer.

That damaging admission was sufficient for the judge. He ordered the plant shut.

The Jacksonville incinerator was in considerable trouble even before Judge Reasoner's decision. On January 23, 1993, an Arkansas state official had revealed that the incinerator was producing a larger volume of hazardous waste than it was destroying. At the end of a year's burning, 9,600 drums of waste had been "destroyed" by the incinerator, but in the process the incinerator had created 12,000 drums of salt and another 1,730 drums of ash, for a net gain of 43 percent in the volume of waste. Furthermore, the salt and the ash are so laced with dioxin that they are legally a "hazardous waste" and thus cannot be taken off the site. In August 1995, the Arkansas Department of Health and Agency for Toxic Substances and Disease Registry (ATSDR) released the results of a study that showed that the dioxin levels in the blood of Jacksonville residents had increased 22 percent since the start of the burning (Costner, 1995a).

—*Peter Montague, Ph.D.*
Excerpted from Rachel's Hazardous Waste News, *#311 and #325*

ell of the Southern Coalition on Protecting the Environment (SCOPE).

Organizing is the Only Way

Sometimes the EPA does use its power to force a corporation to produce less pollution. Sometimes lawsuits are successful in making a corporation pay for some of the harm it has done. Exposés in the media can result in a public official or a corporation adopting new policies or programs. But unless citizens are engaged in an ongoing process of demanding democratic accountability, these changes won't last.

Organizing to rebuild democracy is the only way we can reclaim our health. But successful organizing isn't easy. As historian Lawrence Goodwyn writes in his book, *The Populist Moment*:

Large scale movements happen when they're organized. They happen no other way. And the reason they're not organized more often is that large-scale movements are agonizingly difficult to put together. The entire culture of a society is arrayed against the idea of large-scale collective assertion.

Remembering how difficult organizing is should help us be forgiving of each other and ourselves. If it were easy, it would have already been done. We are trying to build a democratic society without enough blueprints and models, so our trial-and-error method has to leave room for experimentation and mistakes.

Remembering how necessary organizing is should help us be inclusive and persistent. There are no magic facts. There are no perfect heroes to give perfect speeches that will convince the polluters to stop polluting. There is only the dogged determination of people working together to protect their own health, their families' health, and the health of their communities.

What will a "large-scale collective assertion" against dioxin exposure look like? While we can't describe the whole picture, we can identify some of its features. It will be community-based but nationally linked. It will be comprised of people who have been changed by their involvement, and who are figuring out the connections between issues and accepting the centrality of activism in their lives. It will be a diverse, inclusive, celebrating, persistent, experimenting, vibrant movement of people who practice democracy. Out of this practice will come an insistence on the democratic control of the institutions that affect their lives. In *Who Will Tell the People*, William Greider writes:

> If there is a mystical chord in democracy, it probably revolves around the notion that unexpected music can resonate from politics when people are pursuing questions larger than self. I have seen that ennobling effect in people many, many times —expressed by those who found themselves engaged in genuine acts of democratic expression, who claimed their right to define the larger destiny of their community, their nations.

To stop dioxin exposure we must build groups that teach themselves to sing this unexpected music.

Contributors

Charlotte Brody
John Gayusky
Stephen Lester, M.S.
Peter Montague, Ph.D

March and rally to close down the Bronx-Lebanon incinerator, May 8, 1993.
© Clark Jones

Chapter Eleven

People Changing to Change the World

To organize against dioxin exposure, you need the truth. You need a large and diverse group of people. You also need a plan: a clear and simple set of actions that will stop dioxin formation, discharges, and exposure.

This set of actions needs to be built on an understanding of how people change through social action. For an organizing campaign to be successful, two groups of people must change: The people who become involved in the campaign, and the people with the power to make the decisions to stop the poisoning.

People Change As They Become Active

At the April 1995 CCHW Roundtable on Dioxin, participants created timelines on the components of change in their own lives. (See Table 11-1.) Your campaign to stop dioxin exposure needs to give participants the opportunity to go through all the stages of personal change that come with social activism. To be successful, your campaign must also allow different people to be at different stages at the same time. Some participants will be educating themselves; some will be soul-searching, and exploring their feelings of betrayal by a government that won't protect them;

Table 11-1

Stages of Change

- Rude awakening, spark, threat
- Reacting to injustice; anger, fear
- Educating yourself, becoming aware
- Figuring out that you're not alone
- Realizing authority (government, corporations) won't protect you
- Soul-searching, moral outrage
- Acting on love of community, love of country, faith in God, belief in principles, spiritual connection to earth and life
- Accepting that we are the solution; taking action, personal responsibility
- Building confidence; getting encouragement from family and others
- Learning lessons from and being validated by other movements
- Finding role models, mentors, community elders
- Gaining inspiration from victories, seeing the power in people
- Obtaining power over decisionmaking
- Making connection to other issues
- Figuring out that activism is your life

some will be looking for role models and lessons from other movements, and some ready to make a lifetime commitment to social change.

Many organizations start falling apart when they start expecting everyone to be at the same stage of personal change. It is much more realistic to build your group on the understanding that some participants will be ready to take action and win victories while other people will still be trying to understand the problem. Don't stop educating because some people don't need any more information. And don't feel that you have to wait to create direct actions until every possible participant in your community is ready to take that step. Use a committee structure to let people participate at a level at which they are comfortable.

Figure 11-1
Pollution Comes from the Board of Directors

To stop dioxin formation, discharges, and exposure, we must constantly recruit new people. Our efforts must help participants move from education to action to victory and from individual anger to group strength. Activist campaigns must have role models. The spiritual importance of personal and social change must be recognized and honored.

Corporate Decisionmakers Must Change, Too

CCHW Dioxin Roundtable participants described their activism as the outcome of soul-searching and outrage. They took action because they could no longer bear inaction. Similarly, the people who make decisions inside corporations will stop making dioxin when they can no longer bear the responsibility for poisoning people, animals, and our environment.

Right now, the handful of people who make decisions at the chlorine-producing corporations, at the PVC plastics plants, at the chlorine-bleaching paper mills, and at the incinerator companies have decided that dioxin's damage to our health is acceptable to them. The campaign against dioxin exposure must convince these decisionmakers that the harm caused by dioxin is unacceptable.

Corporate decisionmakers who change will experience many of the same stages of change that activists experience. The corporate CEOs, board members, and other decisionmakers will first experience a rude awakening, spark, or threat. Then they will react to the evidence of injustice the campaign is heaping upon them with anger and fear. As you keep providing them with educational materials, they will become aware—although not necessarily accepting—of the facts you've given them.

When corporate decisionmakers figure out they're not alone in being targeted by a campaign, they may join with other corporations in fighting back against the campaign, (for example, by giving more money to the corporations' coalition, the Chlorine Chemistry Council). If the campaign keeps the pressure on these decisionmakers, in time they should realize that making dioxin is harming their corporation's health. Role models—corporations that have changed—will help them reach this realization.

A change of heart may come when profits become unacceptably low, or when there is an unacceptably high level of tension between the decisionmakers who have become convinced that they should stop making dioxin and those who believe that they should proceed with business as usual. It may come when bad publicity fills an unacceptably large part of each work day, when laws penalize dioxin emissions with harsh mandatory fines, when cleaner technologies show substantially higher profit margins, or when the cost of lawsuits from dioxin victims becomes unacceptably high.

The American people want to help us change corporations. Seventy-three percent of the American people believe that "there is too much power concentrated in the hands of a few big companies." Most people in the United States (53 percent) disagree with the statement, "Business corporations generally strike a fair balance between making profits and serving the public interest" (TMC, 1994).

So if you publicize a true story about a corporation knowingly and needlessly making people sick, most people in the United States will believe that the story is true. And if the public has to choose between corporate power and the right to clean air and safe food and water, the majority will choose our rights. But we lose our audience and our battle if our opposition can counterpunch by telling a true but damaging story about us that most people will believe. If people believe the opposition's story, they won't be able to wholeheartedly support our campaign for justice. The stereotypes used by corporations are all too familiar:

- "They're a bunch of 'I've-got-mine' environmentalists who don't care about workers' jobs or the economic future of our community."

- "It's just a scraggly group of back-to-the-earthers who want us all to live on nuts and berries without the benefits of modern technology."

- "The opponents to our project are left-wing, Godless extremists aimed at destroying the private enterprise system that made our country great."

- "They're all hysterical housewives without any accurate information to support their wild fears and accusations."

- "They're unrealistic, selfish NIMBYs."

- "Their so-called leaders are out of touch with the needs of the community they pretend to represent".

Activists must make a lie of these stereotypes. Our stand on the issues and the faces in our meetings must not affirm corporate stereotypes about us. If your group fits any of these stereotypes, you've got to figure out how to add more components or people to your effort.

The basic message of your campaign should evoke the public's finest affections for the promise of democracy. Think of the

Statue of Liberty. A one-line description of your campaign should evoke cinematic images of the people standing together to protect our health, our children, and our rights. If you're all white, or all professionals, or all short-term residents, you won't be able to evoke the righteous indignation of a broad enough public. And you need that indignation to win.

The issue of dioxin cuts across the interests of many different groups of people, giving you a head start on creating the diversity you need. Vietnam veterans and the American Legion have a stake in dioxin exposure because of Agent Orange. Breast-feeding advocates such as the La Leche League will care because of dioxin in breast milk. Women's health organizations, groups concerned with birth defects, dairy farmers, firefighters, cattle ranchers, real estate agents, Native Americans and others whose diet includes large amounts of fresh water fish, health food store owners and shoppers, environmental task forces of churches, parent-teacher associations, and the workers and neighbors of dioxin-producing industries and incinerators all share a special concern about dioxin. A local Stop Dioxin Exposure Campaign that includes even half of these groups will provide you with impenetrable armor when the corporate public relations people start throwing stereotypes at you.

Contributors

Charlotte Brody
Participants in the CCHW Roundtable on Dioxin

Chapter Twelve

The Basics of Organizing

As you begin the task of organizing to stop dioxin exposure, you may be an experienced activist, or a newcomer to local campaigns to protect community health and environment. You may be an individual who is seeking to organize a group, or you may be a member of an already existing group that is seeking to expand or join with other groups to form a coalition. You may live in a community where the predominant source of dioxin is obvious, or you may not. In all of these cases, the same organizing basics will be necessary for a successful Stop Dioxin Exposure Campaign. This chapter will discuss twelve basic organizing principles.

- Talk and listen.
- Figure out who you should talk and listen to first.
- Create and distribute fact sheets.
- Recruit new members.
- Conduct meetings.
- Create an organizational structure.
- Set goals.
- Identify targets.

- Conduct research.
- Take direct action.
- Target the media.
- Use laws and science to support organizing.

Organizing Principle One: You Talk and You Listen

Three young organizers took a long drive through rural California to meet famed farmworker organizer Cesar Chavez. After their hard and dusty journey, they asked Chavez, "Cesar, how do you organize?" Cesar replied, "Well, first you talk to one person, then you talk to another person, then you talk to another person."

"Yes," they responded impatiently, "but how do you *organize?*" Cesar repeated his answer. "First you talk to one person, then you talk to another, then you talk to another."

If you are one, two, or three individuals without an organization, your next step is to talk to others so you can build an organization. If you already are part of an organization, then your next step is to talk to the people in your organization about initiating a local Stop Dioxin Exposure Campaign.

To build the relationships that will hold an organization or a coalition together you must meet and talk with people, one or two at a time. Talking with people doesn't mean giving people a speech and waiting for their applause. Talking with people means listening to their stories, paying attention to their reactions to what you've told them about dioxin, finding out what connects the dioxin issue to their own past experiences and future aspirations, and noting the other issues they may bring up in the course of the conversation.

I was invited to help an organization in upstate New York that was fighting a solid waste incinerator. Except for them, the group members told me, no one else in their community was interested in the issue. When I asked how much listening the group members did to connect their incinerator fight to other issues people cared about, they said they listened a lot. But in a role play on recruitment, they talked *at* me without ever learning who I was or what I cared about.

That evening, in a bar, I was challenged by the group to show them how to recruit a truck driver who was having a beer and watching television.

Accepting the challenge, I sat next to the truck driver and told him about the group's goals. He was clearly uninterested in working to stop the incinerator. I then asked him what he did care about locally, and what he would change if he had the power to do so. I listened for ten minutes while he talked about potholes and how costly his truck maintenance was because of them. When I collected enough information to feel comfortable, I asked him how bad he thought the potholes would get when hundreds of trucks carrying tons of garbage used those same streets each and every day. He joined the group because I listened to him and identified what he cared about and tied it into the local issue.

People get involved when they feel that they have a self-interest in getting involved, and that by getting involved they can effectively act on that self-interest. If you are prepared to listen to people to identify what they care about, and discuss "pocketbook" issues and "nuisance" problems as well as health and environmental concerns, you should be able to appeal to almost everyone's self-interest. Once you trigger someone's self-interest, then you have to convince that person of the ability of your group or coalition to tackle the issue and win.

Organizing Principle Two: Figure Out Who You Should Talk and Listen to First

Organizing a community around dioxin has great potential for broad participation. But you need to figure out who you are going to talk with first. Who are the people most directly affected in your community? Who are the people for whom you are the best recruiter?

At Love Canal, these were easy questions to answer. I needed to talk to the other people who lived around the canal. They were my neighbors, so I was an appropriate recruiter. They were also the people most affected.

You can work out the answers to these questions in a brainstorming exercise at an early meeting of your group or coalition. First you brainstorm a list of the groups of people whose self-interests are most directly affected. Then you figure out who you know that knows each group of people.

Groups of people most directly affected by dioxin exposure include:

• people who work in a dioxin-producing plant, incinerator, or landfill;

• people who live near a dioxin-producing plant, incinerator, or landfill;

• Vietnam veterans and their families and veterans organizations;

• women with endometriosis and their families;

• breast-feeding mothers and their families, and advocates of breast-feeding;

• cancer survivors and their families;

• nurses and physicians;

• people who have experienced infertility and their families;

• children with birth defects and their families;

• farmers and farmworkers who use pesticides;

• home builders and real estate salespeople;

• grocery and health food store owners;

• dairy farmers and their families;

• firefighters and their families;

- cattle ranchers and their families;
- fishers and their families;
- parent-teacher associations;
- environmental and conservation groups;
- junior and senior high school environmental clubs;
- Girl Scouts and Boy Scouts.

To identify and contact these kinds of groups in your community, it may help to look at local organizations listed in the Yellow Pages or in resource books produced by the United Way or the Chamber of Commerce.

Organizing Principle Three: Create and Distribute Fact Sheets

After you've read this book, talked to some people, and decided to start a local campaign to stop dioxin exposure, your next step is to create a fact sheet. An attractive, easy-to-read, and accurate fact sheet educates the community about the problem of dioxin and ties the issue into their lives. If community members think they are directly affected, they will be likely to join you in your efforts to shut down the sources of dioxin. An effective way to generate community interest is to present the dioxin problem broadly, so that it applies to as many people as possible.

A simple, one-page fact sheet will serve the purpose. You might include:

- the names of the authors and/or your group name, address, and telephone number;
- what dioxin is, where it comes from, and how it accumulates in the body;
- how the community gets exposed to dioxin and dioxin-like chemicals;

Figure 12-1

Sample Factsheet

Dioxin and Dinopolis

Dioxin =

Male Reproductive
System Toxicity:
Reduced sperm count
Abnormal testis
Reduced size of
genital organs
Lower male hormone
levels
Feminization

Female Reproductive
System Toxicity:
Decreased fertility
Inability to maintain
pregnancy
Ovarian dysfunction
Endometriosis

On Fetus:
Birth defects
Alteration in reproductive
system
Decreased sperm count
Reduced fertility
Neurological and
developmental problems
fetal death

On Everyone:
Chloracne
Hirsuitism
Hyperpigmentation
Immune system
suppression
Hormonal changes
Altered glucose response
Altered fat metabolism
Wasting syndrome
Diabetes
Damage to liver, spleen,
thymus and bone marrow
Lung cancer
Stomach cancer
Liver cancer

Did you know that every time you drink a glass of milk or eat a hamburger, you are filling up your body with dioxin? Incinerators, chlorine using pulp and paper mills and chemical manufacturing plants that make chlorine based products are filling up the American people with this poison.

In September, 1994, the United States Environmental Protection Agency released a new draft reassessment on dioxin. In this 2,400 page report the EPA reviewed all of the scientific studies that have been done on dioxin and concluded that the American people are full of dioxin -- the average boy, girl, woman, and man has enough or almost enough dioxin in their body tissues to damage their health.

Dioxin comes out of industrial smokestacks and waste pipes when chlorine is used in industrial processes. The dioxin then gets into the fish in the polluted water and into the cows, chickens, pigs, and lambs that end up as meat and dairy products in our homes.

Here in Dinopolis, the St. Elmo Hospital incinerator downtown and the Shiny White Paper Plant in the East River area are major sources of dioxin pollution.

Dinopolis residents have the right to food that is safe to eat. We have the right to swim and fish in a clean Lake Dino. We have the right to grow gardens in soil that has not been poisoned. We have the right to graze our cattle on land that has not been contaminated by an incinerator or a paper plant. Before we get any sicker from dioxin, we must join together to convince dioxin polluters that they must stop the poisoning of their neighbors.

Join us. Come to a meeting of the still forming Dinopolis Stop Dioxin Coalition on Tuesday, February 3, 1996 at 7:00 p.m. at the Dinopolis Regional Library, 118 Main Street in the first floor meeting room. Learn more about dioxin and what we, as a community, can do to stop it.

The Dinopolis Stop Dioxin Coalition presently includes the East River Neighborhood Association, the Dinopolis La Leche League, the Dinopolis Chapter of Viet Nam Veterans of America, the Dinopolis Valley Environmental Justice Alliance, and Local 421 of the Service Employees International Union (SEIU). Your and your organization are invited to join. For more information about the February 3 meeting or the Coalition call or come by the Environmental Justice Alliance, 147 East Frederick, or call (812) 697-4327.

- types of health effects that may result from exposure to these chemicals; and
- how dioxin is affecting businesses and property values.

    If you already know the local sources of dioxin, you can add:

- where the local dioxin sources are located; and
- who owns the local dioxin sources or who originally contaminated the area with dioxin.

    For a sample fact sheet, see Figure 12-1.

Organizing Principle Four: Recruit Millions, One at a Time

Direct contact with people is the basic building block of organizing. If you don't yet have a group, you need to go person-to-person and meeting-to-meeting in order to build one. If you've already got a group, you need to go person-to-person and meeting-to-meeting in order to build a coalition. As you go person-to-person, people will give you a lot of information. You will get a good idea of the magnitude of the interest in the issue in the community, and you will find other ways to connect to the community's self-interest.

Face-to-face conversations make the difference. If you are building a group, face-to-face conversations with other individuals is the way you get them to attend their first meeting. If you are building a coalition, face-to-face conversations with individuals representing groups is the way to get them to join the coalition. Sending a letter to invite someone to a meeting, or placing a call to leave a message about your meeting, is generally not enough to win someone's support for or participation in your efforts.

All recruiting is a form of door-knocking. If you are trying to organize a neighborhood, the doors line the streets. If you are trying to build a different kind of group or coalition, the doors may be spread all over town and you may need appointments to open them.

Before you try to recruit someone, you have to decide what kind of support you are asking for. Do you want the person to come to a planning meeting? Do you want the person to invite you to talk to his or her group's membership? Do you want the person to recruit his or her group or neighborhood into your coalition?

When I first decided to go door-to-door to talk with my neighbors about Love Canal, I was scared to death. As I approached the first door, I was sure that the person who answered would slam the door in my face. Or that the people who lived in that house would think that I was crazy or just causing trouble. I thought, "Who am I to talk to people about a dumpsite and my sick kids?" I was a high school graduate and a full-time homemaker. The closer I got to the first house, the more frightened I became. When I finally did reach the door, I knocked so lightly that no one heard the knock, not even the dog. I ran home feeling ill, and frustrated that someone more skilled than I wasn't knocking on *my* door.

When my son was admitted to the hospital for a second time, my fear and anger turned into action. I realized, while watching him sleep in the hospital bed, that I was now partly responsible for him being there. As a parent who took her responsibilities seriously, I realized that my fear of a stranger's door was overriding my responsibility to protect my child. When my son was well again I went back to that first door and knocked hard enough to make the dog bark.

As I went from door to door, I found that no one slammed the door in my face. In fact, most people were friendly and openly gave me information and volunteered to help. It turned out that many of my neighbors were hoping, like me, that someone skilled would one day knock on their door.

There are several ways you can make knocking on doors easier.

First, organize what you are going to say. You need to put together a "rap" to use at the door. Successful professionals who canvass for money start their rap with, "*I am...*" "*We are...*" "*This is...*" "*We want...*"

Here's how you can use the same approach:

*"I am...*your name.

*"We are...*a small group of parents who are concerned about dioxin.

*"This is...*a fact sheet or petition about the dioxin problem.

"We want ... you to attend a small meeting next Tuesday at our local fire hall to discuss our concerns about this dioxin problem and what we can do together as a community.

Once you have memorized your "rap," practice it in front of a mirror or try it out on your family or friends.

Second, write down all the information people give you in order to avoid future confusion over who said what and who made what commitments.

Third, before you start, select a date and find a place to meet with those who want to discuss the problem. Create a flyer listing the time and place of the meeting, or simply add the information to your fact sheet.

Fourth, consider circulating a petition. You don't need a lawyer to write the petition, you just need to write something simple. For example: *"We the undersigned residents of Our Town, U.S.A., petition the local/state government body/ Chemikill Industries to test the area/explore other ways to manage dioxin-contaminated wastes."* The purpose of the petition is two-fold: to get the names and addresses of community supporters whom you can later contact, and to show community support to those in power.

When you knock on a new person's door, there are those awkward first few seconds when he or she is deciding whether or not to slam the door and go on with the business of life. Your opening has to be clear, straightforward, and appealing. The person will be wondering who you are, where you come from, what you want, and how much what you are asking for is going to cost them. Think about your own experiences with people coming to your door. What makes you decide to talk to them? What makes you decide to close your door?

The questions of who you are, and whether you are connected to anyone the person knows, are credibility questions a good organizer will work out in advance. If you can say, "I was just talking to your neighbor, Mrs. Jones, and she said you'd be a good person to talk to," or "Reverend Smith is working with us—he's letting us use the church basement for our meeting next week," you have borrowed credibility and will have a few more seconds to get in the door and on with your rap.

Whether you are selling brushes or anti-dioxin campaigns, the time comes when you have to close the sale. In organizing, the sale is a commitment from the person to do something. Use your judgment to gauge what the person can afford to do, and remember that everyone can do something. Commitment should be expressed as action. "I believe" should flow directly into "I will do." Just signing the petition is probably too easy. So is making a half-hearted promise to come to the meeting. Try to get a firm commitment to attend the meeting. Explore other ways that the person can become actively involved. Ask for other contacts. Invite the person to knock on doors with you or to make two to five contacts. Make sure the person knows you'll be in touch to see how things went. End the visit when you are sure you and the new person have a clear and concrete understanding about the deal—about how many people the person will contact and how many he or she will commit to bringing to the next meeting.

Before you leave, the new person should know how happy you are with the meeting and, more importantly, how essential he or she is to building the organization. Never forget to say thank you.

If your door-knocking is not door-to-door but all over town, the same basic principles apply. But you will probably need an appointment to knock on those doors. People are more likely to agree to set up an appointment if they know the person who is asking for their time. So if you don't know the person you want to talk to face-to-face, try to find someone who does. Ask this person to make the appointment for the two of you and go together.

Organizing Principle Five:
Hold Meetings That Make People Want to Come Back and Bring Their Friends

According to Tim Sampson, a long-time teacher of community organizing, people will come to a meeting if:

- they have made a commitment to come,
- they have a role or responsibility in the meeting,
- they have an immediate and specific self-interest in the work of the organization, and
- they have past, positive experiences with similar meetings.

To have a successful meeting, your recruitment efforts must satisfy the first and third of Sampson's conditions. The second and fourth condition will depend on how you run the meeting. Use different kinds of meetings to suit different purposes.

House Meetings

This is the kind of meeting many groups hold when they are first forming. The meeting is held at a member's home and the style is informal. One of the biggest benefits of this kind of meeting is the greater comfort level among members.

A good house meeting should last about ninety minutes and have a three-part agenda, with each part taking half an hour. "What is the problem we're facing" is the first part. "What can we do about this problem" is the second. In the third half-hour, you figure out what exactly needs to be done before the next meeting, and who will do what.

Holding house meetings in different locations throughout the neighborhood is a good way to start establishing a community organization. Rotating facilitators of house meetings is a good way to help identify and build leadership skills in your organization.

Planning Meetings

Before your organization holds a general membership meeting or makes any major decisions, you should hold a planning meeting. Leaders and other key decisionmakers within the organization get together at the planning meeting to set agendas, review the work that has been done, and plan activities. When things go wrong in a campaign, nine out of ten times the cause is either poor planning or no planning.

The setting for a planning meeting is less important than the people who are invited to attend. If your organization hasn't set rules on who attends such meetings, think about who needs to come in order to make the activity you're planning a success. Some people should be invited because you can count on their good ideas; others should be invited because their participation gives them a positive role in the process and a sense of ownership. One important addendum is to consider those who may cause trouble if they are not involved in the planning process. You will have to decide whether you want to deal with these people in advance, at the planning meeting, or later, when they raise a ruckus.

Planning meetings should not be decisionmaking meetings. Instead, planning meetings should establish the agenda and a process by which decisions will be made at a general membership meeting, or should define a plan to carry out an activity that has already been decided upon by the membership.

Groups sometimes use planning meetings to make decisions on behalf of the membership. The general meeting then becomes the place where those decisions are announced and ratified. These groups are not practicing democracy. We all know what it's like when the government or a corporation tells us what they plan to do without our consent or permission. Why would we want to replicate the same actions we are fighting against?

General Membership Meetings

Most organizations need to hold regular membership meetings. These ensure that all members of the organization share the responsibility for decisionmaking and for carrying out the activities of the organization. These meetings, however, are the hardest to carry out in a lively and productive way.

General membership meetings should always start with an agenda. The highlights of that agenda should be shared with everyone who is asked to attend. The time and location of the meeting should be chosen to accommodate the maximum number of people. Watch for time conflicts with work schedules, religious, sports or community events, popular television programming, and other previously scheduled events.

Running a good general membership meeting requires good instincts and common sense. You will also need a good sense of balance in order to do the following:

- Make the meeting fun and sociable without seeming silly or frivolous.
- Make the meeting orderly but not stiff.
- Allow everyone to have their say while avoiding long, repetitive speeches.
- Make sure decisions get made efficiently without jamming them down people's throats.
- End the meeting on time while covering all key items in the agenda.

Good planning and shared responsibility are probably the best ways to ensure the proper balance on these matters. The best way to measure your success in holding meetings is by counting how many people come back.

People will come to the next meeting if they enjoyed the first one, if it started and ended on time and wasn't a bore, if it produced concrete results, if it was lively and exciting, and if it delivered what was promised.

Another simple but crucial point: People come to the next meeting only if they know when and where it is. At every meeting, you should announce the date for the next meeting and hand out flyers for people to hang on their refrigerators as reminders. You should also pass out a sign-up sheet to get people's names and telephone numbers so they can be called and reminded of the next meeting or be updated on recent developments on the issue, or contacted for followup calls.

No matter what kind of meeting you hold, you should follow up with everyone who attended. This is one reason why you need a sign-up sheet for attendance. By following up, you guarantee that everyone has a common understanding of what happened and feels that his or her presence was important. Followup gives shy people a chance to talk so that they may feel more comfortable speaking at the next meeting. Also, people who have additional ideas can express them during followup calls.

Sometimes followup can be uncomfortable, especially if the meeting did not go according to plan. All the more reason to follow up! This way, you get views on what went wrong, and you can encourage members to take some responsibility for making the next meeting better.

Organizing Principle Six: Structure Matters

C CHW staff people constantly hear the same three complaints from local leaders:

"I feel like I have to do all the work. Nobody helps me and I'm tired and I want to just quit, but I can't."

"Nobody responds. They just sit there. Even when I tell them what to do, they don't do it."

"The three of us on the executive committee are getting on each other's nerves. Everybody wants to pull a power play. We're not getting anything done except fighting with each other."

All of these problems have their roots in organizational structure. To avoid these problems, take the time to seriously plan out your group's structure. If you already have a group, take a good, close look at your group's structure and make the changes that are needed.

How much structure do you need? Enough to effectively involve your members so that they feel needed and are capable of making decisions—and enough so that you and the other core group members aren't doing all the work. The structure of your organization should help encourage people to join, get active, and stay active. People quit when they feel useless. They also sometimes quit if they are asked to do things that either are too much for them to handle or are too vague or undirected, leaving them feeling that they don't know what they're supposed to be doing.

Most organizations use a pyramid structure, as shown in Figure 12-2. This structure is very efficient for decisionmaking, since decisions are made by a few leaders at the very top. But it does not encourage participation or ongoing involvement from the general membership. Occasionally, leaders of a pyramid will ask the general membership to make a decision. The general membership, unaccustomed to being asked and without the direct experience needed to really decide, may just sit there like couch potatoes, confirming the top leadership's impression that for most of the members, "the lights are on but nobody's home."

Leaders often complain that after six months of a campaign, "only a handful of us are left to do the work. Nobody's coming to meetings." These problems are the cost of a top-down decision-making structure that does not respect or involve the abilities of its membership.

The opposite extreme is a freeform, leaderless structure where decisions are made only by consensus by everyone who happens to be in the room. The assumption is that everybody is at the same stage of leadership. Making decisions becomes an agonizing process. Every meeting starts from scratch, rebuilding consensus. If the opposition decides to infiltrate a consensus-based organization, it can block any action it doesn't want.

A model that compromises between these two extremes comes from Citizens Against Toxic Sites (CATS) of New Castle, Pennsylvania. The wheel structure, as shown in Figure 12-3, is similar to the structure we used at Love Canal.

When new people join the organization, they are asked to join one of several committees. The committee provides them with a specific task that they know how to do and like to do. Delegating tasks this way spreads the work across the organization and

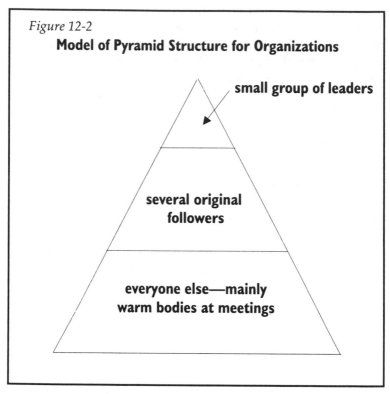

Figure 12-2
Model of Pyramid Structure for Organizations

small group of leaders

several original
followers

everyone else—mainly
warm bodies at meetings

prevents burn-out. Each committee has a general "mission" and is empowered to set up subcommittees if it needs to. (For example, the Public Relations Committee has a squad of folks that does the CATS newsletter, another that coordinates the speakers' bureau, others that produce flyers, etc.) Committees are coordinated by the Executive Committee, which is comprised of two delegates from each committee, plus two co-chairs elected by the general membership. At Executive Committee meetings, all the committees report in on their activities, compare notes, discuss where committee efforts are overlapping, and get clarification or technical approval from the full Executive Committee. Approval for major committee issues takes place at general membership meetings, which give everybody a share in "owning" the organization and provide direction for each committee.

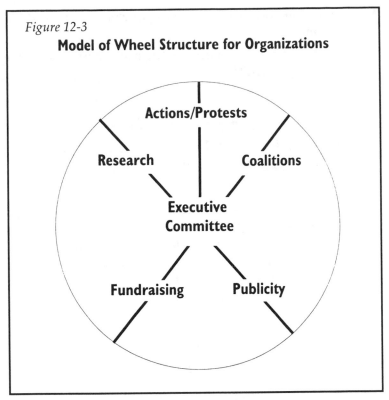

Figure 12-3
Model of Wheel Structure for Organizations

One method Love Canal residents used to get and keep people involved was to appoint block captains who served as "spokes" on the wheel. People volunteered to liaison with a block of ten homes to inform those neighbors about what the group was doing, encourage them to attend membership meetings, and gather their general feelings related to campaign issues. Each block captain's responsibilities included:

- contact by phone or by visit every two weeks,
- distribution of flyers and newsletters,
- contact within a week of a meeting held by the group or the opponents,
- contact about events and fundraisers,

- solicitation of volunteer time and selling or buying tickets to events, and

- attendance at block captain meetings to report back to the coordinating committee.

Organizing Principle Seven:
Set Goals So Everyone Knows Where You're Going and How Far You've Come

People will come out more often and stay with a group longer if they see that the group is achieving its goals. Thus it is critically important to have long-term, intermediate-term, and short-term goals to help members understand where they are going and the steps they have mastered along the way. It is also critical to acknowledge and celebrate your successes no matter how small you may think they are.

The goal of CCHW's Stop Dioxin Exposure Campaign is a sustainable society in which there is no dioxin formation, discharge, or exposure. Progress towards these goals can be measured by reduction in the dioxin in our food and in breast milk.

A long-term goal for a community campaign to stop dioxin exposure could be to achieve a dioxin-free community or to sustain the community without dioxin-producing industries.

Intermediate goals toward these long-term goals could include closing the local medical waste incinerator, getting the local cement kiln to stop using waste fuel in its processes, or convincing the local university to adopt procurement policies that mandate the purchase of only chlorine-free recycled paper.

A short-term goal could be to secure a meeting with the head of the hospital board to discuss the incinerator, or to get a television story produced on the danger posed by the cement kiln. Short-term goals might also include establishing a strong power base of members with active working committees, or having representatives of your organization speak at ten other organizations in the next two months.

The following exercise may help your group decide on its goals:

- Ask people to tell you what they think the goals of the organization should be. List these suggested goals on a large sheet of paper at the front of the room.

- Ask the group to narrow down the list. One way to do this is to break the group into smaller groups, asking each to choose the five goals it thinks are most important to work on. Compile a new list from each small group's list. This new list will be significantly shorter than the original list.

- Discuss the new, shortened list, asking members to explain why they believe a certain goal should stay on the list. A two-minute time limit should be placed on people's explanations. You may want to establish a rule that each person can only speak once until all have had an opportunity to speak.

- Ask for a show of hands or use a written ballot to select the group's top five choices.

For example, your group's top goals might come out like this:

- Goal #1: Create a dioxin-discharge-free zone.

- Goal #2: Close the local medical waste incinerator at our county hospital.

- Goal #3: Stop the use of waste fuel in local cement kilns.

- Goal #4: Pass a law which mandates that local government institutions must purchase chlorine-free paper.

- Goal #5: Get the community's pizza restaurant to stop using white chlorine-bleached cartons.

It's important for everyone to define and mutually agree upon the organization's goals. When you use an exercise like this one to set those goals, the less essential goals will fall to the bottom without making the people who suggested them feel left out. You will also gain insight as to what people care most about. After you've prioritized your list you can separate the goals into long-term, intermediate-term, and short-term. Then you'll know where to start.

Organizing Principle Eight: Identify Targets

Pinpoint the actions and the people that have the power to help you reach each goal. The people who impede the achievement of your goal are often referred to as the targets of the campaign. This does not mean that they are evil or bad. It simply means that because they have the power to give you what you want, it makes sense to focus your attention and actions on them.

The target of the campaign must always be a person or persons. You can't fight City Hall because City Hall is a building, and a building can't decide to stop producing dioxin. But you can target the person with the power at City Hall to act on your behalf. Individual decisionmakers have human responses such as guilt, fear, ambition, vanity, and concern over the public's perceptions of them. These human qualities mean they are capable of being convinced to change. Institutions don't have these human qualities and thus don't offer these opportunities.

To make a list of targets, your group must answer three key questions:

- Who is responsible for the situation you want to change?

- Who can make the change you want happen?

- How can you convince them to act on your issue?

The first question will lead to a list of names of the corporate polluters and the government agencies who have given or can give a "permit to pollute." But instead of just getting the names of corporations, you'll need to find out who owns the corporation, and who makes the decisions about the situation you want changed. For government agencies, you also need to identify who has the decisionmaking power.

The second question will establish a list of government officials ranging from local government representatives to the President of the United States. The list will also include all the people you named in the answer to the first question, because the responsible party or parties can also provide the solutions.

The third step is to figure out how these people—these government decisionmakers and corporate executives—are vulnerable. What would make them do what you want them to do? Some ideas on how corporate decisionmakers change can be found in Chapter Eleven, "People Changing to Change the World."

Organizing Principle Nine: Research is an Essential Tool

Back in 1921, John Brophy, president of the United Mine Workers Union, said:

> Research is digging facts. Digging facts is as hard as mining coal. It means blowing them out, butting them, picking them, shoveling them, loading them, pushing them to the surface, weighing them and then turning them loose on the public for fuel, light and heat. Facts make a fire which cannot be put out.

Research is a tool, not an end product. You need to do research to gather enough information to achieve your goals, not to know absolutely everything there is to know. (Remember: The truth won't set you free, but organizing will.)

Research is necessary to find out your local sources of dioxin. Research should identify and focus on the responsible parties. It should tell you who has the power to give you what you want.

Research can also help you figure out what arguments your targets will probably use against you. Once you know this, you need to do research to create counterarguments. *Is* it more expensive to make chlorine-free paper? If we close down the hazardous waste incinerator, *will* it destroy the town's economy?

Well-researched information is needed to dispel fear and misinformation. If a corporation threatens to move its factory, it is important for you to know whether their threat is legitimate or not. Research could uncover whether the corporation is heavily invested in your town, and whether it might actually go bankrupt if it tried to leave.

Occidental Chemical in Niagara Falls, New York, often said it would leave if the Love Canal residents continued to impose unreasonable demands on the company, or if the workers began contract negotiations. But residents' and workers' research showed that Occidental had just invested millions of dollars in the plant and in a new solid waste transfer and disposal facility. Without that research, both the residents and workers might have believed the company's threats.

Research can help you identify where your opponent is getting money to support its work. A county hospital, for example, receives state and county subsidies. Sometimes a private corporation will receive subsidies in the form of tax breaks, or in reduced or free utility and sewer fees. Or an incinerator may receive public bond money or loans from a local bank. Your group may be able to impact those who have the authority to approve these funds. You can also research a company's history of environmental violations, and labor law violations, and other legal actions against them.

In Palmer, Massachusetts, the Ware River Preservation Society looked into the history of Waste Management (now called WMX), which planned to build a regional landfill nearby. Through its research, the group found that Waste Management had been forbidden from operating in Chicago, Illinois, because of a local "bad boy ordinance." (For a sample of this kind of ordinance, see Appendix C.) This local law prohibited permits from being issued to companies such as Waste Management that had been found guilty of felony violations. The group applied this research creatively and produced a flyer that asked: "If Chicago, the home of notorious gangster Al Capone, decided that Waste

Management was too corrupt for their city, why would our community want them?" The flyers were distributed to every home. The following week, after hearing from outraged local residents, the county commission voted to deny the Waste Management proposal.

You should also research to identify where you might find potential members. After you find out where your local dioxin sources are, you can do a "community analysis" of which sectors of the community have a special reason to join your group. They might live or work near the dioxin source, or have children enrolled at a local school that is downwind or downstream.

When people uncover facts, they "own" them—it builds their determination and will to win. Divide the work of collecting research among your group. Find out who in your group loves the information superhighway, and let them do research online. Find out who loves the library, and assign them research in the library stacks. Who in your group knows the stock market and can read corporate annual

Creating a Plan of Action

Let's say that you want to get the county commissioners to vote against a proposal to build a medical waste incinerator. The commissioners have refused to discuss the issue or vote on it at a regularly scheduled commissioners' meeting, although you know there have been many back room discussions. Several commissioners are up for re-election and your group feels this is the opportunity to force a vote to see where these commissioners stand on the issue. The following could be your group's plan of action:

What is your goal?

To stop the incinerator.

Who is your target?

The commissioners in general, but the chair specifically because he controls the meeting and he is up for re-election.

What are your demands?

To get the commissioners to agree to put the issue on next month's agenda for a full vote.

What will you do?

Get your membership to pack the room at the next

regularly scheduled county commission meeting.

How will you measure success?

If the commissioners agree to place the issue on the agenda; if enough people show up to fill the room; if people walk away feeling good about the activity; if there is some media coverage that will educate the larger public about the issue and the evasiveness of the commissioners during an election campaign season.

What is your theme?

Democracy. Adults will have lap signs that say "In a democracy, we vote." Children will hold up American flags and signs that describe what they have learned about representative government. Someone can dress like Uncle Sam.

You need to plan out every step and every detail and figure out all the things that could go wrong and what your group will do in each situation. Will you have a press conference before the commissioners' meeting? Who will announce your demands? Will people all march in at the same time or will

reports? Is there an accountant or bookkeeper to read corporate tax returns? Are there good "people persons" who can find out all the gossip on who makes what decisions at the incinerator or hospital?

Organizing Principle Ten: Take Direct Actions

In his 1963 *Letter from a Birmingham Jail*, Martin Luther King, Jr., wrote:

You may well ask, "Why direct action? Why sit-ins, marches, etc.? Isn't negotiation a better path? You are exactly right in your call for negotiation. Indeed, this is the purpose of direct action. Nonviolent direct action seeks to create such a crisis and establish such creative tension that a community that has constantly refused to negotiate is forced to confront the issue. It seeks so to dramatize the issue that it can no longer be ignored.

An action is any step you take to advance your group's

goals. Petitions, letter-writing campaigns, and educational meetings are all actions that advance your group's goals. A direct action is the most dramatic type of action, involving confrontation and demands.

You should begin your campaign against a target with the most courteous actions. Your group could write a letter to ask him or her to come to a meeting of your group. Or your letter could ask your target to set up an appointment with key group leaders. If your polite request results in a meeting at which your information and organization convince the target to do what you want done, that is great! Celebrate! But in all likelihood, these actions won't convince your target, so you will need to intensify your actions. Make sure you have copies of everything you've sent. This allows you to prove to the media and to disbelievers that you have tried the more polite approaches.

folks trickle in as they arrive? Who will speak to the commissioners and when? Do they have a sign-up requirement for speaking at these meetings? Who will find out what that requirement is, and who will be sure that key people in your group have their names listed? Who will be responsible for press calls before the event and for producing the press packets? Who will be in charge of the on-site media coordination? Who can talk with the security guards or police if they show up? Who will show people where the nearest public restrooms are? Who will map out the area where your action will take place? Where is the door to the commissioners' meeting room? How do people get there? Who will produce written directions? Is parking available? What other on-site logistics do you need to consider? Role playing can help you to prepare for "best case" and "worst case" scenarios.

Direct action begins after your efforts at education, information-sharing, and persuasion are ignored. Or when every effort you've made has resulted in the creation of yet another corporate or government committee or study group that will take forever to make recommendations that will eventually be ignored.

Use direct action when your group is ready to confront a decisionmaker with its frustrations and to make specific demands. Direct actions move your organization outside the established rules for meetings and discussion. It takes your group into a forum in which you make the rules and where elected representatives and corporate executives are less sure of themselves and of how to handle the situation. A direct action often provides the necessary pressure that forces your target to act on your group's issue.

You must be careful not to overuse or burn out the membership with too many actions. Each action must be carefully planned, with consideration given to such questions as: What do we want to accomplish with this action? Who is our target and are we asking the right decisionmaker for something he or she can give us? Should we take this action now or wait for a better opportunity? Is our action something that our members are comfortable participating in? An action should be a milestone in the life of every participant, so you should consider how the action will feel, whether it will be empowering, and whether it will be fun.

Your group needs to develop a detailed plan of action. One example of an action plan can be found in the sidebar.

Civil Disobedience

Civil disobedience can be a very powerful form of direct action. It works when it honestly reflects your group's and your community's frustration with business and politics as usual. Civil disobedience should be considered when simpler forms of communication and direct action have been ignored.

Acts of civil disobedience are intended to disrupt the corporate or political process that you want to change. Sometimes civil disobedience involves breaking the law your group is trying to change. That's what North Carolina A & T students Ezell Blair, Jr., David Richmond, Franklin McCain, and Joseph McNeill did when they sat down at the Woolworth's lunch counter in Greensboro, North Carolina in 1960.

Other times civil disobedience involves breaking the law to protest a legal practice that is wrong. In 1982, when Warren

County, North Carolina, residents lay down in the road to stop the dumping of hazardous waste in their community, they were practicing this form of civil disobedience. So were the activists at the Boston Tea Party.

Civil disobedience requires even more careful planning than other forms of direct action. How will the police respond? How can you control the situation so no one gets hurt? Where will arrested people be taken? Who will deal with the arraignments? Do participants understand the risks, and are they prepared for the legal consequences of an arrest?

Organizing Principle Eleven: Target the Media

The news media are managed and owned by corporations. To convince radio stations, journals and magazines, newspapers, and television programs to cover your story you have to answer the following two questions: Who are the media's decisionmakers who need to be convinced that our story should be covered? What will it take to convince them that our story should be covered?

In most media outlets, the decisionmakers are the editors, and the way you get them to cover you is to spoon-feed them a story they can use without much work.

To spoon-feed the media a story they can use, you must do the following:

- Know their deadlines.
- Know their areas of expertise.
- Visit them.
- Reward them with scoops.
- Realize that they don't always read their mail, and never bring press packets with them.
- Figure out the formula they use for their stories and use it to write your own story.

- Write your own press release as if it were their story and then be flattered, not offended, when they run the story as their own.
- Hold your events where they can find you.
- Make your events short, creative, and highly visual. Television crews especially hate to cover events that are just a series of talking heads. Also try to see that your events are filled with people similar to the media outlet's audience.
- Never lie.
- Never tell half-truths.
- Provide stories that will help propel reporters into larger media markets.
- Provide a local angle for regional media.
- Practice your sound bytes.
- Always have a spokesperson available.
- Make sure the media know who the spokesperson is and how he or she can be reached at all times.

Television news editors want visual drama that is understandable in ten-second soundbytes. Talk show editors want good talkers that make their telephones ring. Newspaper editors want to win the Pulitzer Prize. None of these editors wants to spend fifteen minutes finding the telephone number you forgot to put on your press release. Few of these editors will assign a reporter to cover a story that is completely new to them.

Sometimes this means that you have to educate the media before they will cover your story. As with every other kind of communication, the best way to do this is in person. Use your door-knocking skills. To get an introduction, find out if someone in your group goes to church with an editor. Does someone have children in the same school? Then take along your fact sheet. Very few editors want to get in trouble with their bosses or their advertising departments, so every fact you provide has to be true. If you're targeting a corporation that advertises in the newspaper or on the television program or radio station, point that out to the editor or reporter you're working with.

Be persistent. The people have a right to know, and they won't know about your story if the press won't report it. Follow up every press release with a telephone call. Make sure that your press release was received by the right person. Ask if the station or paper plans to send a reporter or camera crew. If the answer is no, offer to set up an interview at another time.

If no reporters show up at your event, call the editors and ask them why. Offer to report the story on the phone. Ask reporters and editors for advice on how you can get better coverage next time. Have a few other group members call to ask why there was no coverage.

It is essential to constantly refine and develop your campaign's media strategy. But don't be fooled into believing that the media is the only way to get your story out. Keep creating your own media through fact sheets, cable access television programs, newsletters, call ins to radio shows, letters to the editor, statements at public hearings, barbecues, rallies, auctions, concerts, and videotapes.

New Bedford, Massachusetts

Local activist Angela Days describes New Bedford as "a melting pot of nations." The residents are predominantly non-English-speaking, low-income immigrants. Their community is next to the Acushnet River, which is polluted with dioxin-like PCBs and heavy metals. Their children suffer from leukemia and other contamination-related diseases. Many residents have high levels of PCBs in their blood.

The EPA's "solution" to cleaning up the contamination was to dredge the river, then burn tons of PCB-contaminated sludge in a mobile incinerator. The New Bedford community was determined to stop the EPA's plan, and pleaded with the EPA to reconsider its decision to incinerate. The EPA refused.

Hands Across the River decided to fight the EPA. The group's first step was to get more people involved. Hands Across the River members passed out leaflets in the affected neighborhood inviting residents to a community meeting.

Much to their surprise, attendance at the first community meeting was very poor. Members wondered if anyone really cared. Refusing to be discouraged, however, they walked around the neighborhood and asked residents, face-to-face, why they did not attend the meeting.

Residents told members that they feared the presence of television cameras and reporters, and feared government harassment and deportation. Once an understanding was reached, neighborhood residents proved to be great activists.

Hands Across the River carried out a two-pronged campaign to pressure the EPA: from above by involving congressional representatives, and from below by gaining the support of local government.

Group members inundated their congressional representatives with letters, phone calls, and personal visits. Their strategy was to be "in a legislator's face" as much as possible and demand support in stopping the incinerator project. The group finally received the support it wanted.

Hands Across the River petitioned the local government to let residents vote on a ballot initiative on the incinerator project. The group's strategy

Organizing Principle Twelve: Use the Law and Science to Help You Organize —Not Instead of Organizing

You can use the legal system to fight dioxin exposure, but this method has its limitations. The biggest is that it is not illegal to pollute and to discharge dioxins. Corporations are given permits to pollute by state and federal governments. So you can only use the law when a corporation or the government has done something extraordinarily bad. Even then, community groups that use the legal system need to have plenty of money and time. Establishing legal proof means getting experts to fight their experts, and compiling elaborate documentation of wrongdoing. In Woburn, Massachusetts, $2 million and several years were spent on testing and modeling to prove that twenty-two children developed leukemia from water contaminated by trichloroethylene TCE.

When you find an attorney who is willing to assist you and file a suit, two things usually occur. First, your group loses members, who now believe that the lawyer will save the day. Second, the fight is moved out of the community and into the courts.

Lawsuits can, however, offer certain advantages. One advantage is the possibility of winning financial compensation for health damage, suffering, and lost property values. The operating facility may also be shut down if it is violating the law. Or the lawsuit may delay the granting of a permit, giving you time to organize to stop the facility. The media usually will report on court activity, making your organizing more visible.

Relying on the strength of scientific information rather than organizing can also cause problems. For one thing, it leads to "dueling experts syndrome" instead of campaigns based on common sense. Dueling experts syndrome can start when a local group brings in one scientist who says the community has a dioxin problem. Then the government or dioxin-polluting corporation

was to create an opportunity for residents to express themselves through the voting process; educate the public about the dangers of the project; and win the vote to stop the incinerator project.

The ballot initiative passed with two-thirds of the voters opposing the incinerator project. Now Hands Across the River had the clout to make city and county officials take action to stop the EPA's incinerator plans.

Rumors soon spread that, despite the successful ballot initiative, the incinerator was en route to New Bedford. The New Bedford City Council then took three very important steps that finally ended the project. The council first passed an ordinance that prohibited incineration inside city limits. It passed a second ordinance that placed weight restrictions on transported items within city limits (the incinerator would exceed the weight limit). It then refused to issue water or electricity permits for the incinerator construction site.

Together, New Bedford residents and local officials created impenetrable barriers. The EPA was so outraged it threatened to air drop the incinerator, generator, and water tanks into New Bedford. It also threatened the city with fines of $25,000 a day for delaying the project. These threats, however, were never carried out, and the incinerator never came to New Bedford.

New Bedford residents, having experienced the power of democracy and community, are confident that as the EPA now looks for other cleanup options, residents will play a vital role in determining the future of their community.

brings in two other experts who say there is no problem and no cause for alarm. Members of the community then become confused because they are not sure who is right, and because most people don't understand the technical arguments or the jargon used.

When the dueling experts syndrome occurs at public meetings, the community loses interest. Members have no role to play, they can't engage in the discussions, and they have no say since the fight is now in the boxing ring of science.

In community after community, people have identified and documented health problems associated with dioxin and dioxin-like substances. Studies proved that in Love Canal, 56 percent of the children were born with birth defects; that in western Oregon, there was a thirteen-fold increase in the number of infants born with neural tube birth defects; and that on Long Island, New York, the closer a woman lives to a chemical plant the greater her likelihood is of having breast cancer. None of these terrible findings alone convinced the dioxin polluters to stop polluting. Nor did these documented problems—directly correlated to contamination—convince the EPA to step in and force the responsible corporations to protect the community.

However, there are several advantages to using science as a part of your strategy to stop dioxin exposure. First, the information provided by your scientists will give your group credibility. Scientists like Theo Colborn (whose new book on dioxin will be out in February 1996) have helped communities understand the science behind their problems (Colborn, 1996). Their studies can show that you are not fabricating these horrible stories. Second, science can provide information on the types of health problems people experience as a result of dioxin exposures. This information may help your group identify others in your community who might be interested in joining. Scientific information showing dioxin levels in cow's milk and meat, for example, can trigger a group effort to recruit ranchers or dairy farmers.

History has shown us that the decisions to evacuate a community, clean up a dumpsite, or force an existing industry to clean up its dioxin discharges are political decisions. They are brought about by organizing, not by lawsuits or the power of scientific data.

The first evacuation at Love Canal was a result of the pressure Love Canal residents exerted on New York Governor Hugh Carey while he was running for re-election. Love Canal residents followed Carey everywhere, carrying signs and distributing fact sheets and press statements. Residents held the governor personally responsible for the Love Canal situation.

However, political strategies need to be supported by scientific information and sometimes by legal maneuvers. You can use the scientific information you have gathered to put pressure on your target. As Love Canal residents followed Governor Carey around, we said over and over again, "56 percent of our children have been born with birth defects. How many more children must be born deformed before you act to protect them?"

Contributors

Charlotte Brody

Residents of Woodlawn, North Carolina, protest the siting of a toxic waste in-cinerator directly across from their homes, September 14, 1991. © *Jerome Friar*

Chapter Thirteen

Building a Coalition

Before we can stop dioxin exposure and rebuild democracy, we will have to become much more powerful than we are now. The kind of power we need comes from building a movement; from engaging community after community in demanding and practicing self government. To begin this process, we have to expand beyond the individuals and groups with whom we usually work. This means creating coalitions with other groups. To gather the strength we need to win this campaign, we need to sit down at the table with veterans organizations, dairy farmers, breast-feeding advocates, trade unionists, sport and subsistence fishers, firefighters, and other groups with whom we may never have worked before.

Working in coalition can be very difficult. But the factors that make coalition work uncomfortable also make it important. The people and groups in a coalition don't always see the world the way you do. They dress differently. They talk about different things. They don't do what you do on Saturday night. The fact that you are working together to stop dioxin exposure in spite of your many differences allows you to create strategies that will resonate with more people, proving to the outside world that something very important must be going on.

In order to build on their strengths and avoid unnecessary conflicts, member groups must recognize that each organization will bring its own history, structure, agenda, values, culture, leadership, and relationships to a coalition. In *Environmental Poli-*

tics: Lessons from the Grassroots, North Carolina activist/researcher Bob Hall writes about building a multiracial group.

> To make a multi-racial alliance work takes (1) constant energy, negotiation, education, and commitment; (2) a self consciousness among the leaders of their limitations without, and strengths with, a coalition; (3) consistent delivery on promises made and holding up one's end of the bargain; (4) a recognition of differences, including sometimes conflicting agendas; (5) a recognition of the power of racism in the history and contemporary life of the community and beyond; (6) education of the membership about the need of multiracial partnership; and (7) lots of practical steps that aim to solidify personal and political relationships. (Hall, 1988)

The efficacy of a coalition of ten groups with few or no active and committed members is defined by the equation $10 \times 0 = 0$. However, the effectiveness of a coalition that consists of strong organizations with active membership bases can be defined by the equation $10 \times 1 = 50$. That is, a successful coalition's power can be greater than the sum of its parts because the coalition's composition legitimizes the issue and highlights its importance to the community.

In a coalition to stop dioxin exposure, each member group will have some organizational self-interest in dioxin and its related effects on health, jobs, the economy, and/or the environment. Member organizations will bring different strengths and weaknesses to the coalition. As long as each member understands and accepts what other members bring, the coalition should encounter minimal conflicts. For example, one organization may be able to contribute large sums of money but be unable to turn out its membership for an action, while the opposite may be true of another member organization. Both contributions are essential to the success of the coalition's efforts and both groups should be valued by coalition members for the resources they bring.

The Advantages of Coalitions

A coalition of organizations working together can win on more fronts than a single organization working alone.

A single group may be able to run a successful campaign against a single dioxin polluter. But to truly eliminate exposures to dioxin, we need to do more than remove a single dioxin-producing site. We need a lot more people with more contacts, skills, and energy.

A coalition can develop new leaders. As experienced group leaders move up to lead the coalition, they leave openings for new leaders in each of the individual groups. With training, the new leadership will emerge, broadening the leadership ranks.

A coalition will increase the impact of each organization's effort. Involvement in a coalition means there are more people who have a better understanding of your issues and more people who can advocate for your group in some way.

A coalition will increase available resources. Your group can directly benefit from additional members, and may also share resources with other coalition members, such as office space, meeting space, printing equipment, or money. Your group will be able to access all of the contacts, connections, and relationships these other groups have established.

A coalition will broaden your organization's scope. A coalition may provide the opportunity for your group to work on regional, state, or national issues, making the results of your local efforts more far-reaching and effective.

A coalition can build a lasting base for change. When groups unite, each group's vision of social justice and collective social change broadens, thus weakening opponents' ability to label the coalition's efforts as those of "special interests."

A successful coalition is comprised of people who have never worked together before. Coming from diverse backgrounds and different viewpoints, they have to figure out how to respect each other's differences and get something big accomplished. They have to figure out how each group and its representatives

Wake and Moore Counties, North Carolina

"It's all about empowerment and community control," explains Nathaneete Mayo, Chair of the Shiloh Coalition.

Two dioxin-contaminated Superfund sites in North Carolina are the Koppers Company site, located in an African-American community in Wake County, and the Aberdeen Pesticide site in Moore County. Though they are located in two distinct communities with different contaminated sites, the Shiloh Coalition in Wake County and MooreFORCE in Moore County are working together to share information and join forces to create a larger "people power base." Their sharing comes easily since they are experiencing the same problems and living through similar struggles.

Both communities have massive chemical contamination, including dioxin, in their soil and ground water. Both communities relied on groundwater for their drinking water supplies, and both have lost community and individual wells. In addition, the residents in both communities drank this contami-

can make its different but valuable contribution to the overall strategy for change. Isn't that putting the promise of democracy into action?

The Disadvantages of Coalitions

One danger of working in coalition is that member groups can get distracted from other work. If that happens, not only will the non-coalition efforts be less successful, but the coalition will be weakened as well. Your group needs to identify ways to make the dioxin issue relevant to each group's other work, to prevent the coalition from becoming an excessive distraction.

The coalition may only be as strong as its weakest link. Different member organizations will have different levels of resources and experience and will grapple with different internal problems. Organizations providing a lot of leadership and resources can get impatient with some other groups' inability to deliver on commitments.

But coalitions require many compromises. To keep the coalition together it is often necessary to cater to the lowest common denominator, especially when negotiating tactics. Groups that prefer more confrontational, highly visible tactics may find that the more subdued coalition tactics are not exciting enough to activate their members.

In coalitions, the range of experience, resources, and power among member groups can create internal problems. The democratic principle of one group–one vote does not always sit well with groups with a wide range of power and resources. Your coalition needs to carefully define the relationships between powerful and less powerful groups.

Individual organizations may not get credit for their input in a coalition. If all activities are done in the name of the coalition, groups that contribute a lot often feel they do not get enough credit.

So how do we emphasize the advantages and minimize the disadvantages of coalitions? By acknowledging the importance of recruitment, by encouraging the growth of

nated water for years before the chemical contamination was discovered.

The two groups have worked together to win cleanups at both sites. Neither community was satisfied with original cleanup proposals, and both fought to have safer, alternative cleanup technologies used. In Wake County, the groups are looking at a dechlorination process as a substitute to a mobile incinerator. In Moore County, a thermal desorption process is being investigated as a substitute to a mobile incineration plan.

The Shiloh Coalition and MooreFORCE have been very successful in organizing their communities, and finding ways to connect and work together. They are also principled in their view of the contamination in their communities. They will not allow their wastes to be dumped on another unsuspecting community. The groups fought removal plans first in Wake County, where they derailed a proposal to burn the waste in an incinerator in Kansas (Newman, 1994).

new leaders, by defining the coalition's areas of agreement, by taking time to learn about each group, by getting the right people to the table, by giving recognition to individual groups, by attending other member groups' events but never getting involved in the internal politics of other member groups, and by creating a decisionmaking structure that works.

Recruitment

To recruit groups into your coalition, you need to pinpoint all the possible organizations that have a stake in stopping dioxin exposure, identify the right people to talk to in each of these groups, and then go door-knocking. (See Chapter Twelve, "The Basics of Organizing.")

Once your group has decided which other groups it might approach, the next step is to prioritize the list of groups. Which one of these groups will need to take just a small step to join the coalition? Which of these groups would have to take a big leap to join the coalition?

Figure out which people in your group are best to visit the most likely coalition members. Does anyone in your group know anyone in these organizations? Get a clear commitment from your group members on who they're willing to visit. Decide and role play what you are going to say in these visits.

Remember that you need to offer the group you are asking to join your coalition something that members see as in their self-interest. As organizer Tim Sampson once said, "The flowers of organizational relationships grow from personal interest, kindness, and cultivation."

Pick a date by which all first-round visits will be done. It may help to first choose a meeting date on which the first round of prospective coalition members can meet. Put together a timeline that allows you time to visit each group at least a few weeks before the initial coalition meeting date.

Go organization door-knocking. Make sure each person that visits an organization reports the results of that visit back to a central coordinator.

Hold the meeting with prospective coalition members. You could begin the meeting by having everyone introduce themselves and then discuss the issue of dioxin in your community. Distribute a proposed statement of purpose for the coalition and a suggested description of expectations for member organizations. Then encourage feedback from other groups.

The best meetings are those with a clear set of questions to be answered, and an established process that lets everyone at the meeting have a say in answering those questions. At the end of the meeting you should have an agreed-upon set of operating principles that each representative can take back to his or her group.

Labor and Religious Organizations

Religious institutions and labor organizations are two established kinds of groups that are essential to your coalition. They offer people, resources, power, and credibility.

Stopping dioxin exposure and rebuilding democracy can be seen as spiritual issues. The involvement of religious institutions in the coalition can help to focus the entire efforts of the campaign back to the fundamental values of stewardship, compassion, and justice. Most communities have several councils of religious leaders such as an Interfaith Council, a Black Ministerial Alliance, an Evangelical Ministers Association, or the Council of Churches. These groups can help map out the landscape of perspectives in the religious community and suggest who in the religious community is most likely to join your coalition.

Most churches have adult study groups or Sunday school classes in which current issues are discussed. Some churches also have environmental working groups. If you don't know people in these groups, you can make an appointment to talk with the clergy person in charge of that church, synagogue, temple, or mosque.

The history of labor unions is one of struggle against corporations and for democracy. This history can be an essential component of your efforts. The involvement of unions will also help to build strategies that target a corporation's decisionmakers, not the workers on the shop floor. Labor organizations also play an active role in local electoral politics by endorsing candidates and working in electoral campaigns.

In the fight against dioxin, unions and other coalition members can work together to build a safe workplace and environment. In West Virginia, a community coalition is working to assure that a proposed paper mill is chlorine-free, and also that it is built and run with union labor.

We should not expect workers in dioxin-producing industries to agree to sacrifice their jobs for the cause any more than we would expect other coalition members to sacrifice theirs. Our demands must include job protections for all affected workers. When we have the power to convince corporations to stop producing dioxin, we can use that power to protect workers' livelihoods as well.

Identifying Areas of Agreement

It is rare that all members of a coalition agree on all issues. You need to agree to disagree on some issues, and stay focused on the one issue—dioxin—on which you do agree. If the disagreements are so fundamental that they interfere with all of your work, then working together just may not be possible. You need to flag fundamental disagreements immediately before they infect the entire coalition.

One successful method of establishing areas of agreement is to list them. Let participants suggest areas of agreement. Go around the room and see if each member organization agrees with each proposal. Then let people suggest disagreements. Again, go around the room to check if these are actual areas of disagreement. Finally, try and get a consensus from the members to place the list of disagreements aside for the sake of this effort.

Getting the Right People to the Table

Member organizations' choice of representatives to the coalition board is an indication of how seriously they take the coalition. If the presidents of member organizations do not themselves participate, they should send a high-ranking board member or staff person as their regular representative. Some coalitions make participation dependent upon sending an "important" single representative. Most important is having the same people at the table each time. Having a series of individuals "fill in" for the named representative on the coalition can be very disruptive.

Working arrangements depend upon the size of the coalition. For example, in a large national coalition, and even in statewide coalitions in large states, it is very cumbersome and expensive to bring the whole board together regularly. Such coalitions often choose a working group or committee to develop strategy. As long as the board is clear on how policy is set and trusts the working group to confer with them at appropriate points, this method can be useful. Whatever the structure, it should be one that is clear to all coalition group members.

Recognizing Resources

One of the most frequent problems in coalitions involves the giving and receiving of credit. The fighting and jockeying over who gets recognition for what often seems petty. Some may feel that this is something that groups need to be cured of, and that the proper attitudes will make it go away. But, quite to the contrary, these problems are rooted in a basic survival instinct. They will never go away, nor should they.

An organization's ability to raise money, recruit members, build power, attract staff, develop leaders, and fulfill its mission depends directly on the amount of public credit it receives, particularly in the media. Coalitions that lose sight of "giving credit where credit is due" don't last long.

When the issue of the coalition is of secondary importance to a particular member group, then the issue of giving credit is less of a problem. But when the issue of the coalition is also the main issue of the member group, then the issue of giving credit is a thorny one. The coalition's strategy needs to be structured so that there are actions the affiliates do jointly as a coalition, and others that the coalition helps member groups do in their own names. For example, a statewide study of dioxin sources produced by the coalition can be released to local papers by the local member groups who helped conduct it, under their own names. At the same time, the study can be released to the statewide media under the name of the coalition. Groups join coalitions to gain power, not to give it away.

Keeping Out of Internal Battles

Never become involved in the internal politics of any other coalition member organization. Stay neutral during member group's internal election campaigns or fights. Be careful not to become involved in jurisdictional fights between unions in your coalition or in turf battles between community groups. At the same time, try to remain aware of these internal struggles. If you are unaware of these problems, it is easy to fall into the trap of having the coalition take an action that favors one side or the other.

A Sample Coalition Structure

The following section provides possible examples of how an effective coalition might be created and sustained. The procedure for joining the coalition, for example, could be that the coalition is presented with a letter signed by two officers of the member organization, indicating that the organization has voted to join the coalition. The coalition could then vote on accepting the new organization.

The coalition's expectations of member organizations could be that each organization will elect representatives to the coalition's board. At least one of these representatives will attend each monthly meeting of the coalition, and serve as voting delegates. Each member organization agrees to educate its membership on dioxin, to take an active role in educating the community, to pay $50 in annual dues, to participate in fundraising events, and to take part in strategic activities endorsed by the coalition.

The decisionmaking process could mandate that no activity will be undertaken by the coalition unless 60 percent of the voting delegates at any regularly called meeting vote to endorse the activity. (This decisionmaking process should be used for all proposed activities. There should be no side deals. No one should ever feel pressured to take part in any other coalition member groups' activities unless it has come up for a vote and passed.)

Individuals should be encouraged to join the coalition and participate in all activities. Individual dues, for example could be $15 per family per year. Individuals will be considered members when they have signed a membership statement committing themselves to the goals of the coalition and paid their dues. A voting delegate will be elected to represent every fifty individual members of the coalition. Individuals may be elected to office in the coalition.

A coalition may elect officers and committee chairs at any regularly scheduled meeting. A majority vote by the voting delegates present is necessary for election. Spokespeople for the coalition may be approved in the same manner. Nominations may be taken from the floor for all offices.

The coalition should have clear goals, monthly timelines, and people responsible for each agreed-upon activity. Everyone should agree to report to the designated coordinator of each project or activity.

All of the questions surrounding money are critical to understanding who has control over the coalition. Reports on income and expenditures should be a part of every coalition meeting.

If certain groups feel excluded or do not participate, member groups need to understand why in order to better understand the politics and self-interest surrounding the issue. However you design your coalition, make sure you take the time to provide it a structure and a framework in which the different groups can learn about each other and from each other.*

Contributors

Charlotte Brody

---

*Several ideas in this chapter came from *Organizing for Social Change: A Manual for Activists in the 1990s* (Bobo, 1991).

Chapter Fourteen

How To Be a Dioxin Detective

The first step to stopping dioxin exposure is finding the dioxin sources in your community. But since no chemical company or utility is in the business of making dioxin, the sources won't be listed in the Yellow Pages under "dioxin—manufacturers" or "dioxin—retailers."

How do you start to make a list of dioxin sources in your community? Dioxin is generated from industrial chlorine chemistry during production, use, or disposal. You will have to look for all the places where chlorine is "cooked" or where the wastes from cooking chlorine are dumped. Below are five resources you can use to uncover the sources of dioxin in your community:

- *The knowledge of your group.* Write up a list of the types of dioxin sources described later in this chapter and hand it out to members. Make copies of this list of major sources (see Table 14-1). Give everyone a chance to think about operating smokestacks and production facilities in your community, and then share what they know about these facilities. Put every known source on a list that everyone can see. Make a separate list of the questionable sources that need further investigation.

- *The local chamber of commerce or the business section in your public or community college library.* Ask to see the industrial directories for your area. In these handy books you can find all the local

Table 14-1
Sources of Dioxin Released into the Environment

- medical waste incinerators
- municipal solid waste incinerators
- hazardous waste incinerators
- cement kilns
- pulp and paper mills
- chemical manufacturing
- production of PVC, synthetic pesticides, chlorinated solvents
- wood burning
- metal smelting and refining
- wastewater treatment plants
- coal burning
- motor vehicle fuel
- chlorine gas production
- petroleum refining
- forest fires
- tire burning
- recycling scrap electrical wire
- drum and barrel reclamation
- carbon regeneration furnaces
- burning PCBs and PBBs
- burning residential oil and gas
- residential incinerators in high-rise apartment buildings
- backyard burners
- industrial furnaces and boilers especially those that burn hazardous waste oil

manufacturers of chlorine, the producers of plastics and pesticides, the pulp and paper mills, the smelters, and other possible dioxin-producers.

- *The permitters.* Incinerators and energy plants should have air pollution permits. A list of the permit holders should be available from the permitting agency in your state. Call your state environmental agency or regional EPA office to start.

- *Citizens Clearinghouse for Hazardous Waste.* On May 18, 1995, the Center for the Biology of Natural Systems (CBNS) published an important study, *Quantitative Estimates of the Entry of Dioxins, Furans, and Hexachlorobenzines into the Great Lakes From Airborne and Waterborne Sources.* To complete this study, the Center's team of researchers compiled a national list of cement kilns, hazardous waste incinerators, sewage sludge incinerators, hospital incinerators, and iron and copper smelters. The

CBNS has provided CCHW with this list, and we'd be happy to provide you with information on your area. We also want to know what other dioxin emitters you've identified.

- *The Right-to-Know Network.* The Right-to-Know Network carries Toxic Release Inventory (TRI) data, Records of Decision (ROD), and the National Priority List (NPL) of Superfund sites. The TRI provides information on chemicals released from industrial plants into the surrounding community. You can use the TRI data available from this on line service to find out if your local pesticide or plastics manufacturer is releasing chlorine or a chlorine compound. (Look for chemical names containing "chlor".) Also available are "Form Rs", forms that the EPA requires companies to submit every year. These forms describe the facility, where it sends the waste off-site, and information on the chemicals that are released. The ROD (EPA's description of what methods they will use to clean up a site and

Newark, New Jersey

In June 1983, dioxin contamination was discovered in Newark, New Jersey. Diamond Alkali Company (Diamond Shamrock) was responsible for contamination at the abandoned pesticides plant. The three and one-half-acre site borders the Passaic River in a densely populated urban area, and contamination was found both on and off the site. Dioxin levels as high as 51,000 parts per billion were found along a city street near the site.

The Ironbound Committee Against Toxic Waste (ICATW) was formed by local residents who were fearful of adverse health affects from the pollution. They organized the community and fought hard to have the site cleaned up and to obtain some health testing to determine the effects the dioxin had on their health. They won an initial phase of cleanup, but this still left dioxin contamination in the community, and residents are still uncertain whether the cleanup has safeguarded them from further dioxin exposures.

Residents' fears about the toxicity of the site were confirmed as they watched New Jersey Department of Environmental Protection (NJDEP) employees conducting the cleanup at the abandoned plant. The workers wore hoods, gloves, boots, and suits which they would take off at the end of the day and seal in fifty-five-gallon containers. Seventy-nine thousand cubic yards of dioxin-contaminated materials were collected from the site and encapsulated. Groundwater from the site was pumped and treated.

By 1986, the cleanup of off-site contamination was completed as well. In all, 980 shipping containers, 2,000 drums and 80 tanks full of dioxin-contaminated materials were stockpiled, along with the on-site waste collected earlier, on the Diamond Shamrock property. While workers wore moonsuits as they cleaned up the areas surrounding the site, neighborhood children ran around barefoot. People watched, feeling unprotected, unwarned, and angry that the community was given no safeguards against additional dioxin exposure.

In March of 1990 new testing found that fish (both fin

why) and NPL information on your local contaminated site documents the contaminants found at the site, including if dioxin is present. CCHW and many state and grassroots organizations already have an account on the Right-to-Know Network, so you can get on-line immediately through these groups. To open an account with the Right-to-Know Network, call (202) 234-8494. You can also sometimes get this information from people who work in the plant, or by requesting material safety data sheets on the regulated chemicals stored or used on-site from your local emergency planning committee or the local fire department.

After you've identified as many sources as you can, create a dioxin map. Maps are a powerful education tool. Use a city or county road map, the local planning commission's maps, or the 7.5 minute topographic quadrangle maps from the United States Geological Survey to record your information on. The Geological Survey maps show pipelines, railroad tracks, forests, streams, transmission lines, and other features that may not be on a road map. If your covered area is an entire city or county, you will need several of these maps. The Geological Survey produces an index map for each state that makes it easier to figure out which map(s) you'll need.

Drafting supply stores often carry Geological Survey maps and may carry other base maps as well. Camping stores sometimes carry Geological Survey maps as well, but may only have them for popular hiking areas. You can order maps and an index map from the National Cartographic Information Center Geological Survey at (703) 648-6892.

Once you've compiled your list, and made a map of your community's dioxin sources, send CCHW a copy of your list and map with your name and address. Together we can create a national map of dioxin sources.

fish such as striped bass and shellfish) in Newark and Raritan Bays and in the Atlantic Ocean off Sandy Hook (twenty miles south) had the highest dioxin levels ever detected in any food. ICATW believes this contamination is a direct result of the years of discharges from Diamond Shamrock.

No followup testing has been done around the site to determine dioxin levels. Cleanup of the polluted waterways wasn't scheduled to begin until 1995. The dioxin wastes from the cleanup activities still remain, in barrels, tanks, and containers, on the old Diamond Shamrock site.

Some residents and workers have received a million-dollar out-of-court settlement. The judge found, "Diamond knowingly and routinely discharged contaminants over a period of 18 years." The company's waste disposal policy essentially amounted to "dumping everything" into the Passaic River.

—*Arnold Cohen (Newman, 1994)*

Sources of Dioxin Exposure

Once you have identified your local sources of dioxin exposure, you need to do some research on each source. The detailed information in the rest of this chapter will get you started. Feel free to read only those sections of this chapter that describe dioxin sources in your community. Then you can find out the other things you need to know, like who owns each source.

Medical Waste Incinerators

The amount of dioxin and dioxin-like chemicals released into the air from medical waste incinerators in hospitals and laboratories dwarfs all other sources of dioxin. Burning medical waste releases large amounts of dioxin for two reasons. The first is the large amount of chlorine in the waste stream, which comes mostly from polyvinyl chloride (PVC) plastic. The second is the fact that because most of these incinerators are small, emissions are not well controlled. Few have air pollution control devices such as electrostatic precipitators, scrubbers, or fabric filters.

The EPA estimates that there are 6,700 medical waste incinerators in the United States. This includes hospitals, laboratories, veterinary hospitals, nursing homes, animal shelters, and funeral homes. The American Hospital Association (AHA) argues that the EPA's estimate is too high and that there are only about 2,300 medical waste incinerators (AHA, 1995). However, spot-check calls to several state permitting agencies identified many more medical waste incinerators than were listed by the AHA (Cohen, 1995a). The true number is likely somewhere in between the AHA's claim and the EPA's estimate.

To estimate the total yearly air emissions of dioxin from medical waste incineration, the EPA documented the amount of medical waste produced in the United States, then measured dioxin emissions at six of the 6,700 medical incineration facilities. A "medium" confidence rating was assigned to the estimate of the amount of medical waste, and a "low" confidence rating to the level of emissions. The range is 1,600 to 16,000 grams TEQ per year, with a geometric mean of 5,100 grams TEQ. According to the EPA, this estimate of dioxin emissions from medical waste incineration accounts for 53 percent of total known dioxin emissions to the air. (See Chapter Three, "Where Dioxin Comes From.")

The AHA has also challenged these emissions estimates, arguing that the amount of medical waste generated is less, that many facilities do have air pollution control devices, and that emissions are no more than 150 grams per year (AHA, 1995). In response to the AHA's challenges, the EPA has acknowledged that its emissions estimate may be high, but not substantially so. According to the EPA, medical waste incinerators still remain the greatest single source of dioxin emissions in the United States.

Municipal Solid Waste Incinerators

Municipal solid waste incinerators are the second largest source of dioxin in the United States. The EPA estimates that between 170 and 190 municipal waste incinerators operate in thirty-seven different states, burning about 17 percent of all solid waste or household garbage generated. (The precise number is unknown because some facilities shut down temporarily and later reopen.) The states with the most incinerators are New York (sixteen), Florida and Minnesota (fourteen), Massachusetts and Virginia (eight), and Connecticut (seven). The locations of municipal solid waste incinerators in the United States are shown in Figure 14-1.

There are three different types of municipal waste incinerators: mass burn (the most common type), modular, and refuse-derived-fuel (RDF) incinerators. Each may have different air pollution control devices, the most common being electrostatic precipitators (ESP), dry scrubbers, and fabric filters, or some combination of these. Electrostatic precipitators remove large dust particles, scrubbers remove acid gases, and fabric filters (also referred to as "baghouse" filters) remove small respirable dust

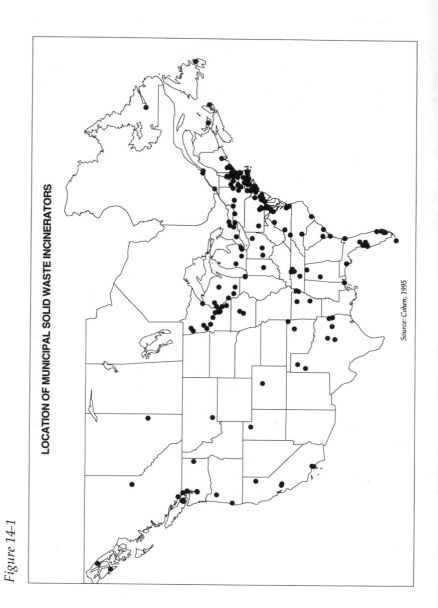

Figure 14-1

LOCATION OF MUNICIPAL SOLID WASTE INCINERATORS

Source: Cohen, 1995

particles. Twenty-seven of the estimated 170 incinerators in the United States do not have any pollution control devices.

Each type of incinerator, with or without air pollution control devices, results in the release of different amounts of dioxin. When the EPA conducted thirty tests of different types of municipal waste incinerators, they found that two types of incinerators account for most of the dioxin emissions. Mass burn incinerators with electrostatic precipitators burn 11 percent of the garbage waste burned in the United States but release 72 percent of the total dioxin emissions from all municipal incinerators. RDF incinerators with electrostatic precipitators burn 17 percent of total U.S. garbage that is incinerated and release 24 percent of the total dioxin emitted from incinerators. Together, they release 96 percent of the dioxin from solid waste incinerators, even though collectively they burn only 28 percent of the total U.S. garbage that is burned (USEPA, 1993). Electrostatic precipitators are the reason for this. ESPs operate in a temperature range that is ideal for the formation of dioxin.

The EPA estimated emissions from municipal incinerators to be from 1,300 to 6,700 grams TEQ per year, with a geometric mean of 3,000 grams, assigning the estimate a medium confidence rating (see Chapter Three for a discussion of this rating system).

This estimate, which accounts for 31 percent of the total known dioxin emissions to the air, may be too low. A single garbage incinerator with an electrostatic precipitator in Columbus, Ohio, was found to release 984 grams TEQ per year (OEPA, 1994), almost one-third of the EPA's estimate for all solid waste incinerators combined (OEPA, 1994). A second test from this same incinerator, conducted by consultants for the city of Columbus, found only 200 grams TEQ per year, but this is still a significant contribution from a single source. The accuracy of this second test has been questioned and an investigation of wrong doing is underway (Fitrakis, 1994; Sanjour, 1994a, 1994b). In any case, the testing at this single incinerator raises questions about the accuracy of EPA's estimate.

Garbage incinerators also generate another dioxin-contaminated by-product—ash. Two types of ash are generated by incinerators: (1) Bottom ash or what's left in the bottom of the burner after incineration is complete, and (2) fly ash, the particles collected by the air pollution control equipment. This ash is usually disposed of in garbage landfills, which allow the ash to mix with the numerous leaching agents (such as organic solvents) typically found in these landfills. As a result, dioxin, once trapped in ash, can become mobile again and be re-released into the environment. In some instances, ash is disposed in specially designed "monofills" that only accept ash from incinerators. These landfills may initially contain the ash, but in the long run they suffer the same problems as all landfills and are likely to leak dioxin and other chemicals into the surrounding community (HWN, 1990, 1991).

Hazardous Waste Incinerators

There are 190 hazardous waste incinerators in the United States (see Figure 14-2) that burn between 1 and 1.3 million tons each year. This is only a fraction of the 216 to 250 million metric tons of hazardous waste produced annually (Dempsey, 1993). Hazardous waste contains large amounts of chlorine compounds and dioxin precursors, so it is not surprising that burning hazardous waste releases considerable amounts of dioxin. Based on emissions testing from only 6 of these 190 incinerators, the EPA has estimated dioxin emissions of 35 grams TEQ per year. The EPA has low confidence in this estimate and provides a range of from 11 to 110 grams TEQ per year.

The EPA's estimate of the contribution of airborne dioxin from hazardous waste incinerators may be low, because its estimate of the amount of hazardous waste burned each year may be low. In 1994, the U.S. Government Accounting Office estimated that 5 million tons of hazardous waste are burned in the United States each year (USGAO, 1994). If the GAO estimate is more accurate than the EPA's, then hazardous waste incinerators may generate five times as much airborne dioxin emissions as the EPA has estimated.

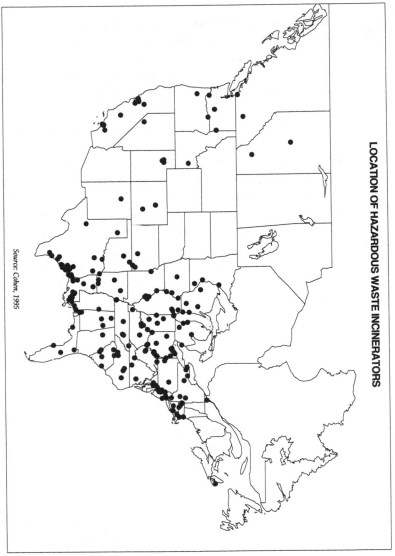

Figure 14-2

LOCATION OF HAZARDOUS WASTE INCINERATORS

Source: Cohen, 1995

Cement Kilns

Cement kilns are used to fuse lime, silica, alumina, and iron into cement clinkers, which are then ground into powder and sold as cement. Most cement kilns use oil or coal to fuel the necessary high temperatures, but some also use hazardous waste, waste oil, wood chips, or tires as auxiliary fuels. Thirty-four of the 212 cement kilns in the United States burn hazardous waste for fuel.

All cement kilns emit some dioxin, but those that burn hazardous waste emit eight times as much as those that do not. In addition, dust from the kilns contains some dioxin, which may be put back into the kiln or landfilled. Some of this dust is released into the environment when the wind picks up dust from the kiln operation and from the piles stored on-site. Because so few kilns were tested for dioxin emissions, the EPA's estimate of their emissions was given a low confidence rating. The range is estimated to be from 110 to 1100 grams TEQ per year, with a geometric mean of 350 grams. The locations of cement kilns in the United States are shown in Figure 14-3.

Pulp and Paper Mills

Pulp and paper mills are the third largest known source of dioxin emissions. Most of their emissions are released via wastewater discharges. The wood used to produce paper contains lignin, a dark-colored substance that must be removed in order to produce white paper. Chlorine acts to separate the lignin from the pulp which can then be removed. When chlorine is used as the "bleaching" agent to remove the lignin, it reacts with the lignin to produce dioxin, which then appears in the paper products, and in the wastewater and sludge from the mills (USEPA, 1987). Dioxin-contaminated wastewater runs into bodies of water, where the dioxin concentrates in the aquatic food chain, ending up in fish. The dioxin-contaminated sludge from paper mills, which is mostly landfilled, represents one of the bigger dioxin reservoirs.

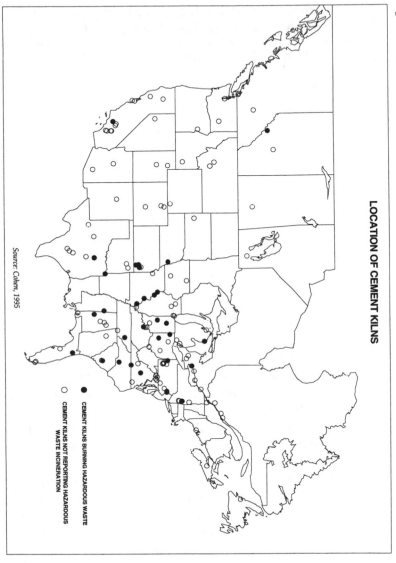

Figure 14-3

LOCATION OF CEMENT KILNS

Source: Cohen, 1995

● CEMENT KILNS BURNING HAZARDOUS WASTE

○ CEMENT KILNS NOT REPORTING HAZARDOUS
WASTE INCINERATION

Table 14-2

Reduction in Dioxin Discharges from Pulp and Paper Mills in Grams of Toxic Equivalents/Year

Discharge	1988	1993
Wastewater	356	105
Sludge	343	100
Pulp	505	150

Source: USEPA, 1993a

Alternatively, some paper mills spread their sludge on land, which poses a significant risk to people and wildlife. Others burn their sludge, which releases very high levels of dioxin into the air.

Dioxin that gets into pulp ends up in paper products. If the paper is used for packaging food products, dioxin can contaminate the food. The U.S. Food and Drug Administration (FDA) found that dioxin migrated into food from coffee filters, cream cartons, orange juice cartons, paper cups used for hot beverages and soup, paper plates used for hot foods, dual ovenable trays, and microwave popcorn bags (LaFleur, 1990). According to the FDA, dioxin levels in these paper products ranged from 1 to 13 parts per trillion (ppt). When these bleached paper products are discarded and burned or buried, the dioxin contained in them can re-enter the environment.

In 1988, the EPA and the pulp and paper industry conducted a survey of dioxin discharges from 104 paper mills. Between 1988 and 1993, the industry reported changes that resulted in a 70 percent reduction in dioxin discharges, as shown in Table 14-2.

According to the EPA, the pulp and paper industry's 70 percent reduction in dioxin emissions was due to "process changes of a pollution control nature." In the past few years, some mills have started using chlorine dioxide instead of chlorine gas (elemental chlorine) in the bleaching process to further reduce dioxin wastewater discharges. But even mills that use only chlorine dioxide still produce dioxin and other organochlorine waste products, although at lower levels (Gruber, 1993). Recent testing by the pulp and paper industry has found that levels of dioxin in wastewater discharges were "nondetectable," but these results have not been published or independently verified. The EPA has not conducted its own testing of dioxin discharges from mills using chlorine dioxide.

The reductions of dioxin in sludge and pulp were not actually measured. Instead, these reductions in dioxin levels were assumed to have occurred in the sludge and pulp to the same degree as in the wastewater. Specific testing of the sludge and pulp needs to be conducted in order to verify that these parallel reductions have in fact occurred. Nevertheless, the EPA has high confidence in the sludge and pulp estimates.

Mills that use recycled paper are not a significant source of dioxin (USEPA, 1993a). Only trace levels of dioxin have been found in wastewater discharges from mills that produce recycled paper or that do not use chlorine bleaching (Berry, 1993). This is because most recycled paper does not need to be "rebleached" with chlorine. Instead, non-chlorine processes such as hydrogen peroxide, ozone, or hot water are used to clean the fibers.

Some mills are moving completely away from using chlorine to bleach paper. The Chlorine Free Trade Association lists 55 pulp and paper mills worldwide that are Totally Chlorine Free (TCF). Although only one of these plants is located in the U.S. (Louisiana-Pacific in Samoa, California), at least another 13 U.S. mills use chlorine-free processes during pulp production. These additional mills, however, add non-TCF pulp to their final paper products (CFTA, 1995). The Canadian provinces of Ontario and British Columbia have enacted legislation that mandates that pulp mills eliminate organochlorine discharges by 2002 (Greenpeace, 1995).

Chemical Manufacturing

C hemical manufacturing is a fairly significant source of dioxin in both air and water, but the EPA has not estimated dioxin emissions from this source. This is inexcusable, since the agency had found dioxin at numerous manufacturing sites as part of its National Dioxin Survey (USEPA, 1987). As part of this survey, the EPA identified 100 2,4,5-T production and disposal sites, between 300 and 600 pesticide formulation plants and 67 "other" organic chemical production sites. At these sites, dioxin is released during the manufacturing process, when the manufactured products are used, and when the products are eventually disposed of, particularly if they are incinerated.

In some instances, a company may "flare" gases at its facility in order to get rid of chemical gases. A flare is an open flame atop a tall pipe, through which gases are vented and burned. Burning chlorinated chemicals or dioxin precursors in a flare may result in dioxin emissions. No testing has been done on flares to determine whether they emit dioxin. The chemical and chemical-related industries known to emit significant amounts of dioxin are discussed below.

Production of PVC

In 1990, more than 11 million metric tons of vinyl chloride and ethylene dichloride, the building blocks for making polyvinyl chloride or PVC, were manufactured in the United States (USITC, 1991). The EPA did not estimate dioxin emissions from the plants that manufacture these chemicals, although they did provide a 1993 Greenpeace estimate. Using data from Sweden and other European countries, Greenpeace estimated that dioxin emissions from PVC feedstock chemicals probably amount to between 230 and 450 grams TEQ per year in the United States (Greenpeace, 1993).

In Europe, nine studies have documented dioxin emissions from facilities that produce vinyl chloride or ethylene dichloride. Dioxin was found in air emissions, wastewater discharges, waste tars, and in sediment downstream from the discharge pipes of these facilities (Thornton, 1995). In 1994, the Swedish Environmental Protection Agency found dioxin in concentrations ranging from 0.86 to 8.69 ppt TEQ in the PVC product itself (Swedish EPA, 1994).

Using information from these nine European studies, Greenpeace updated its estimate of dioxin emissions from PVC production facilities. The studies showed that dioxin can be produced at the rate of 400 grams TEQ per 100,000 tons of ethylene dichloride produced. Using a *very* conservative estimate of 5 to 10 grams dioxin TEQ per 100,000 tons ethylene dichloride produced, and assuming U.S. production of 10.4 million tons of ethylene dichloride per year, Greenpeace estimated that from 500 to 1,000 grams TEQ per year of dioxin would be released in the United States from ethylene dichloride production alone (Thornton, 1995).

This estimate puts ethylene dichloride production among the largest sources of dioxin emissions in the country. This is not surprising, since the conditions for manufacturing ethylene dichloride and vinyl chloride are "nearly perfect" for dioxin formation. Chlorine, carbon, and oxygen are present in a thermally and chemically reactive environment. In addition, metal catalysts, which further increase dioxin formation, are present.

Further evidence that dioxin is produced at these types of facilities was provided in 1994, when Greenpeace collected and analyzed samples from nine U.S. chemical plants that produce ethylene dichloride or vinyl chloride. Four samples were analyzed for dioxin and twenty-five were analyzed for other substances used in the production of these chemicals. Greenpeace reported the following findings:

- At Vulcan Chemicals in Geismar, Louisiana, a sample of "heavy end" waste from the distillation of ethylene dichloride contained 200,750 ppb of dioxin.

- At Formosa Plastics in Point Comfort, Texas, a sample of heavy end waste from the distillation of vinyl chloride contained 761 ppb of dioxin.

• At Georgia Gulf Corporation in Plaquemine, Louisiana, a sample collected from a tank containing process waste contained 1,248 ppb of dioxin.

• At Geon Corporation (formerly BF Goodrich) in LaPorte, Texas, a sample collected from sediment downstream from the discharge point contained 2,911 ppt of dioxin.

In the twenty-five samples from these nine facilities, the following chemicals were found: hexachlorobenzene, 1,1,2,3,4,4-hexachloro-1,3-butadiene, tetrachlorobenzene, pentachlorobenzene and 1,1,3,4-tetrachloro-1,3-butadiene. All of these chemicals are considered "signals," of the presence of dioxin (Costner, 1995a).

Dioxin is released not only when PVC plastic is manufactured, but also when it is burned. For this reason—and because of the prevalence of PVC in the waste stream—PVC plastic is associated with virtually all of the largest dioxin sources identified by the EPA, including medical, municipal, and hazardous waste incineration; and steel, copper, and lead smelters. Since PVC production is the largest and fastest growing segment of the chlorine industry, dioxin emissions from this source can be expected to increase.

Synthetic Herbicides and Pesticides

Dioxin is also found as a by-product in the manufacture of a family of chemicals called chlorophenols. The major chlorophenols are 2,4-dichlorophenoxyacetic acid (2,4-D), 2,4,5-trichlorophenoxyacetic acid (2,4,5-T), and pentachlorophenol. Of these, pentachlorophenol and 2,4,5-T are the most heavily contaminated with dioxin. 2,4,5-T was one of the main ingredients of the herbicide known as Agent Orange, used heavily during the Vietnam War. This chemical is no longer manufactured. 2,4-D is a herbicide used widely today (mostly as Chevron Chemical Company's Weed-B-Gon), and its use releases dioxin into the environment.

A large number of pesticides containing chlorine may also contain dioxin. The EPA initiated two "Data Call-Ins" that required manufacturers of these products to test for dioxin and provide the information to the EPA. However, the results are considered confidential business information, so the EPA cannot make it public. A list of pesticides that are known to be contaminated with dioxin is provided in Table 14-3. In addition, the EPA lists ninety-three pesticides, including endosulfan and parathion, that "could be contaminated with dioxin if synthesized under conditions that favor dioxin formation" (USEPA 1994d).

Chlorinated Solvents

The manufacture of chlorinated organic solvents—many of them, such as trichloroethylene, and perchloroethylene ("perc") used as cleaning solvents—accounts for close to 10 percent of all chlorine production (Thornton, 1994). Dioxin is released in the manufacture, use, and disposal of these chlorinated sol-

Table 14-3
Pesticides Likely to Be Contaminated With Dioxin

Bifenox
Choranil
Chlorophenols
2,4-D
2,4 DB Salts
2,4-DP
Dicamba
Dicapthon
Dichlorofenthion
DMPA
Erbon
Hexachlorophene
Isobac
Nitrofen
Pentachlorophenol
Ronnel
Sesone
Silvex
2,4,5-T
2,4,5-trichlorophenol
2,4,6-trichlorophenol
tetradifon

Source: USEPA, 1994d, PICN, 1985

vents, and has been found in perchloroethylene and 1, 2-dichloroethane at levels as high as 50 ppt (Heindl, 1987). Other solvents that have been found to be contaminated with dioxin included epichlorohydrin, carbon tetrachloride, hexachlorobutadiene, and hexachlorobenzene (Rossberg, 1986; Heindl, 1987). The EPA made no estimate of emissions from these possible sources.

Chlorinated Dyes and Pigments

One class of dyes, called dioxazine compounds, is made from chloranil, a chlorinated "ring" compound that is heavily contaminated with dioxin. Dioxin levels as high as 3,000 parts ppb have been found in chloranil. Chloranil is not manufactured in the United States, but is imported in order to manufacture these blue and purple dyes here in the U.S. In 1992, the EPA discovered that dioxin levels in chloranil could be reduced to less than 20 ppb by using different feedstocks and by making specific process changes. In 1993, the EPA negotiated agreements with chloranil importers to switch to low-dioxin chloranil as soon as they deplete current stocks. The EPA plans to issue a new rule requiring the use of low-dioxin chloranil, but this rule may get stymied in the 1995-1996 Congress.

Another class of dyes containing dioxin and PCBs is yellow dyes that are based on the chemicals diarylide yellow or phthalocyanine. The EPA has prohibited the use of these pigments if they contain more than 50 parts per million of PCBs.

The process of making cottons "wash and wear" can leave high levels of dioxin in clothing. New textiles also sometimes contain high quantities of dioxin. One survey found great variability in dioxin levels in textiles. Even in T-shirts of the same color from the same manufacturer, levels ranged from as low as 50 ppt to as high as 290,000 ppt (Horstmann, 1994). This dioxin may come from the pentachlorophenol that is sprayed on cotton bales as a preservative during sea voyages. This practice has been prohibited in the United States since 1987, but is still used by other countries.

Only a small percentage of clothing contains high dioxin levels, but this contributes to dioxin found in household wastewater and sewage sludge. Dioxin is gradually removed from fabric by repeated washings. From clothes, these chemicals enter the water in washing machines, then the sewer system, then sludge, and the sludge is either spread on land or burned. From the land, air, and water, these chemicals get into the food chain and into us. This cycle illustrates one of Barry Commoner's laws of the environment: "Everything goes somewhere."

Dioxin is also found as a by-product in the manufacture of chlorobenzenes which are used in making dyes as well as pesticides, solvents, and other products. There are 107,500 metric tons of chlorobenzene and 63,000 metric tons of dichlorobenzene manufactured in the United States each year. Trichlorobenzene is also produced, but the EPA did not provide quantities. Hexachlorobenzenes are no longer manufactured intentionally, but they are still formed as by-products of chlorobenzene production and chlorine extraction. According to the few measurements made, hexachlorobenzene may contain measurable amounts of dioxin. The EPA did not estimate dioxin emissions from these products.

Wood Burning

According to the EPA report, over 40 million metric tons of wood are burned each year to heat homes. Another 80 million metric tons are burned by industry for heat, primarily by the paper industry using scrap wood.

The EPA estimated dioxin emissions from industry's wood burning to be 320 grams TEQ per year, with a range of 100 to 1000 grams TEQ per year. Because this estimate is based on measurements at only two facilities, the EPA's confidence in this estimate is low. It is possible that the estimate is too high because one of the facilities where measurements were taken used wood that had been stored in ocean water. This wood would have had many more chloride ions (from the salt in the water), so the dioxin emissions would be higher than typical. The EPA estimated dioxin emissions from residential wood burning to be 40 grams TEQ per year, with a range of 13 to 63 grams TEQ per year.

Wood can become contaminated with dioxin or dioxin precursors before it is burned through dioxin fallout from airborne emissions, drift from aerial spraying of pesticides and herbicides, or chemical wood preservatives such as pentachlorophenol. These processes significantly increase the amount of dioxin generated during burning. Because it is practically impossible to find wood that is unaffected by any of these sources, no one knows whether uncontaminated wood would still generate dioxin when burned.

Metal Smelting and Refining

The smelting and refining of iron, aluminum, copper, nickel, and magnesium can release dioxin into the air. Primary smelting and refining plants extract different metals from ore. Recycling or secondary smelting and refining plants use scrap metal as their metal source, and release more dioxin than primary plants do. This is because the sources of scrap metal contain chlorine from plastics and solvents. For example, when copper wire that contains polyvinyl chloride (PVC) coating is used as scrap metal, the PVC coating is burned off and dioxin is released.

At iron recycling plants, scrap iron is dipped in a bath of hydrochloric acid to remove rust. In this bath, the iron and chlorine from the acid combine to form ferric chloride, which is called "scale." When this clean iron, containing ferric chloride scale, is melted down in a steel vat, dioxin is formed. Scrap iron is also recycled at sintering plants where primarily iron dust and scraps from other processes are recovered. Because of the presence of trace amounts of chlorine and organic matter in the dust and scraps, dioxin is formed in these plants as well (Lahl, 1993). The major sources of dioxin from metal refining are iron sintering plants and secondary copper smelting and refining plants. The locations of these plants are shown in Figures 14-4 and 14-5.

According to the EPA, there are twenty-four primary and secondary copper smelters and refineries in the United States. These plants release 234 grams TEQ per year, with a range of 74 to 740 grams TEQ per year. The EPA's confidence in the accuracy of this estimate is low. However, Cohen et al identified only thirteen secondary copper smelters in the United States, and they are very confident that their number is correct (Cohen, 1995).

The EPA did not estimate dioxin emissions from primary and secondary iron refining. Cohen estimated the release from

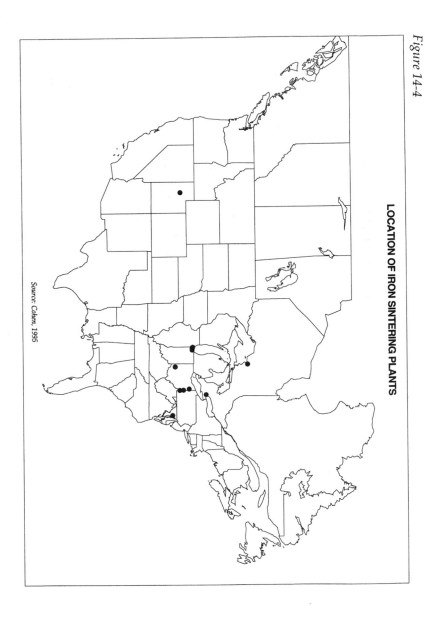

Figure 14-4

LOCATION OF IRON SINTERING PLANTS

Source: Cohen, 1995

Figure 14-5

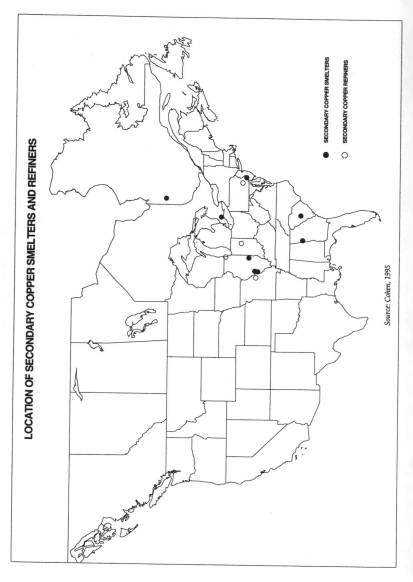

LOCATION OF SECONDARY COPPER SMELTERS AND REFINERS

● SECONDARY COPPER SMELTERS
○ SECONDARY COPPER REFINERS

Source: Cohen, 1995

iron sintering plants to be 230 grams TEQ per year (Cohen, 1995). Almost no measurements have been made of other metal refining and smelting in the United States. Most of the information about dioxin emissions from these sources comes from measurements made in Europe. This industry is one that the EPA should be examining in greater detail.

Aluminum is the major recycled metal in the United States. There are more secondary (recycling) smelters in the United States for aluminum than for any other nonferrous (non-iron-containing) metal. An estimated 1.7 million metric tons of aluminum were produced by secondary smelters in 1987. According to the EPA, no studies have been conducted on dioxin emissions from secondary aluminum smelters in the United States. Testing of stack gasses at a Finnish aluminum recycling plant did find dioxin (Aittola, 1992).

Twenty-three facilities in the United States recycle lead from used car batteries, but, according to EPA, dioxin emissions from these sources are negligible, less than 2 grams TEQ per year. Magnesium refining involves a step in which magnesium chloride is produced by heating the ore and coke in a pure chlorine atmosphere. These conditions of heat, catalyst, and chlorine encourage the formation of dioxin. Measurements at one magnesium plant in Norway showed a release of 500 grams TEQ per year into wastewater and 6 grams TEQ per year into the air (Oehme, 1989). Magnesium plants could be a major source of dioxin in the United States. The EPA did not estimate dioxin emissions from this source.

Nickel is refined by using either a low-temperature process that releases negligible quantities of dioxin, or a high-temperature process that releases much more. However, no measurements of dioxin emissions have been taken at these plants, and EPA did not make any estimate in their report.

Table 14-4

Dioxin Released from the Disposal of Sewage Sludge in Grams of Toxic Equivalents/Year

Disposal Method	Amount of Dioxin Released
Land application	86
Landfill	99
Incineration	23
Surface disposal	20
Sold as fertilizer	4
TOTAL	232

Source: USEPA, 1994d

Wastewater Treatment Plants

Sewage sludge generated by wastewater treatment plants contains dioxin. Possible sources of dioxin in the wastewater stream are industrial and small business discharges into the sewer system, storm runoff that contains dioxin deposited from the air, household wastewater, or chlorination processes in the sewage treatment plants themselves. The EPA has identified household wastewater as a source of dioxin, but it is unclear how important this source is. One study found that almost all the dioxin in household wastewater comes from washing clothes colored with dyes that contain dioxin, wash and wear cottons, and cottons with randomly high dioxin from being sprayed with pesticides (Horstmann, 1994). The EPA provided no information on the role of chlorine bleach in the formation of dioxin in household wastewater. Some evidence indicates that household chlorine bleach is contaminated with dioxin, and contributes to dioxin in household wastewater (Thornton, 1995).

About 5.4 million dry metric tons of sewage sludge are produced each year. The disposal of this sludge releases about 232 grams TEQ per year of dioxin. As shown in Table 14-4, most of this waste is spread on land or landfilled. Although only a small amount of sewage sludge is burned, the 199 sludge incinerators in the United States release an estimated 23 grams per year TEQ of dioxin, a significant amount. Some sewage sludge containing dioxin is sold and marketed as "compost." Some is spread on farmland, which causes surface and groundwater contamination, dust dispersion, and volatilization into the air. Cows or other animals grazing on this farmland can also become contaminated with dioxin, which can come back to us in our food. The dioxin in sewage sludge that is landfilled or burned can also re-enter the environment.

Coal Burning

Coal accounts for 25 percent of energy production in the United States. Dioxin emissions result from coal burning because of chlorine present in coal. No measurements of dioxin emissions from coal burning have been made in the U.S. and the EPA made no estimate in their report. Testing at one European facility showed no dioxin in the emissions, but the procedure used in this test was not very sensitive. Assuming that dioxin levels must be lower than this test was able to detect, the amount of dioxin released from coal burning could be no more than 300 grams TEQ per year. Cohen *et al* estimated dioxin emissions from coal burning to be 200 grams TEQ per year (Cohen, 1995).

Motor Vehicle Fuel

According to EPA estimates, burning gasoline and diesel fuel produces considerable amounts of dioxin. The EPA estimated that 97 percent of the dioxin emissions from motor vehicles came from burning diesel fuel. Burning leaded gasoline, which is no longer used in the United States generates higher dioxin emissions than burning unleaded gasoline. Leaded gasoline contained "scavengers" that include chlorine, such as dichloroethane or pentachlorophenate. The role of these scavengers was to make the lead more volatile and thus easier to escape the engine before it caused damage. In the U.S., leaded gasoline was phased out starting in the mid-1970s to reduce human exposure to lead, although many leaded gasoline cars are still on the road. This action to eliminate lead from gasoline has resulted in an overall decrease in dioxin emissions from automobiles. However, unleaded gasoline remains a source of dioxin emissions because it, too, contains chlorinated chemical additives.

Trucks, buses, and heavy equipment using diesel fuel are the main source of dioxin emissions from motor vehicles. In Great Britain, vehicle emissions are the second largest source of dioxin, after incinerators, primarily because diesel fuel is used so much. The EPA estimates dioxin emissions from motor vehicles burning diesel fuel to be 85 grams TEQ per year, with a range of from 27 to 270 grams TEQ per year and 1.3 TEQ per year (range 0.4 to 4.1) for unleaded gasoline. The EPA's confidence in both estimates is low. The EPA's best estimate of dioxin emissions from burning leaded gasoline ranges from .02 to 19 grams TEQ. U.S. data on leaded gasoline is limited since its ban in 1985.

Chlorine Gas Production

For many years, chlorine gas was produced through electrolysis of salt using graphite electrodes in mercury cells—known as the chloralkali process. This process produces dioxin. Chlorine gas was found to be contaminated with dioxin-like chemicals including PCBs and tetrachlorobenzene (Hutzinger, 1991). The residual sludge left after processing contained high levels of 2,3,7,8-TCDD, and was a major source of dioxin exposure for plant workers. Today, graphite electrodes are no longer used, having been replaced by metal anodes. However, the graphite sludge disposed in landfills represents a potential reservoir of dioxin. In addition, recent tests in Sweden have identified dioxin in waste from non-graphite electrodes (Thornton, 1994).

Petroleum Refining

The production of high-octane gasoline results in some dioxin production. The catalytic reforming process at an oil refinery is conducted at high temperatures and pressure with a catalyst. During the process, coke (a mixture of chlorinated "ring" compounds) deposits on the catalyst. Because the catalyst is expensive, it is regenerated by burning off the coke and being reactivated with chlorine gas. The gases formed during this process contain dioxin and dioxin-like chemicals at reasonably high levels. In addition, the gases may be scrubbed with wash water that picks up dioxin. The total amount of dioxin released from this source was not estimated by the EPA due to insufficient information.

Forest Fires

Forest fires can release dioxin, but as discussed in the Introduction, this is likely due to contamination of trees, leaves, and other forest material with dioxin or dioxin precursors from airborne fallout and pesticide and herbicide applications. Only one study has tried to measure dioxin emissions from this source. The EPA used this study to estimate the emissions from forest fires at 86 grams TEQ per year. The scientists who carried out the study said that they could not distinguish between dioxin created naturally by the burning of wood, and dioxin from anthropogenic (man-made) sources that ended up on forest leaves. The EPA also presented data that has shown dioxin levels from man-made sources exceeding natural sources by more than 100 to 1. Other data has shown that dioxin levels increased sharply with the growth of the chemical industry, and that virtually no dioxin was present in ancient human tissue. This data rules out any significant contribution from the "natural" burning process in forest fires.

Electrical Equipment

PCBs, which are used as a coolant, have been banned from manufacture in the United States since 1977. However, there are still many PCBs in use, especially in electric transformers and capacitors. Electrical equipment contaminated with less than 50 parts per million PCBs does not, by law, have to be replaced. The EPA allows companies to use this equipment until the coolant is no longer effective.

Some PCBs are dioxin-like, and all PCBs can be contaminated with dioxin. When electric transformers and capacitors are disposed of, they may release dioxin and PCBs into the environment. Dioxins and PCBs will be released if the transformers are burned.

Other Sources

There are a number of small sources of dioxin, each of which the EPA estimates as contributing less than 10 grams TEQ per year into the air. Nevertheless, these sources could be significant on a local level.

- *Tire burning.* About 11 percent of the 242 million scrap tires generated each year are burned. Measurements of dioxin released from a tire incinerator with air pollution controls suggest that emissions are in the range of 0.1 to 1 grams TEQ per year. This estimate could be low, because many places that burn tires do not have good air pollution controls. Many tires are burned in municipal waste incinerators, cement kilns, and even paper mills, where they contribute to dioxin emissions from those sources. This source of dioxin could increase as more tires are burned.

- *Recycling scrap electric wire.* Electric wire is covered by insulating material, which can be plastic, burlap, or asbestos-laden fabric. Chlorinated products such as PVC are often used for insulation. In the past, these insulating materials were burned off and the wire sold to a secondary metal smelter, where the metal was reclaimed. The burning of this scrap wire would provide perfect conditions for the formation of dioxin because of the chlorinated plastic insulation, the organic material, and the copper wire, which acts as a catalyst. However, the scrap metal industry says that burning off insulation is no longer an industry practice; instead, the wire is chopped into small pieces and the casing removed by air blowing.

- *Drum and barrel reclamation.* Fifty-five gallon drums and barrels are burned, or in some cases heated, to remove waste material. The cleaned drums are then repaired, relined, and reused. Because these drums contain oils, paints, pesticides, and other chemicals that are dioxin precursors, dioxin will form when these barrels are burned. The EPA estimates that there are twenty-three to twenty-six plants in the United States that burn drums, each handling between 500 and 1,000 drums per day. The total dioxin emissions from this source are estimated to be about 2 grams TEQ per year. The EPA has low confidence in this estimate, which ranges from 0.5 to 5.4 grams TEQ per year.

- *Carbon regeneration.* Granulated activated carbon is used to remove organic chemicals in water treatment systems. Municipal wastewater treatment plants use activated carbon to clean finished drinking water, as do many individual home water filters. Once the carbon is saturated, or full, it is "regenerated" in furnaces by using heat to drive the volatile organic chemicals out of the carbon. The regenerated carbon is then reused. Because of the many dioxin precursors present in the carbon, dioxin and dioxin-like chemicals are formed in furnaces. The EPA estimated that 0.14 grams TEQ per year are released from this source. Its confidence in the estimate was given a medium rating, and the range was 0.06 to 0.3 grams TEQ per year.

- *Burning PCBs and PBBs.* If PCBs are burned accidentally in a fire or deliberately in an incinerator, then considerable amounts of dioxin will be released. Often, in a building fire, the electrical transformers that contain PCBs catch fire. Likewise, the EPA reported that polybrominated biphenyls, widely used as flame retardants in resins, textiles, and paints, can form brominated dioxins and furans when heated under certain conditions. No estimate of emissions from these sources was given by the EPA.

- *Additional combustion sources.* The EPA identified burning residential oil and gas as a potential source of airborne dioxin emissions, but felt it could make no estimates of this source because of the lack of data. In addition, residential incinerators such as those used in high-rise apartment buildings (Laber, 1994), as well as open pit backyard burning, may be other sources of dioxin. Tests done for the state of Illinois found dioxin emissions from a backyard burner to be 2.3 times higher that the level allowed for an incinerator recently permitted by the state (Patrick Engineering, 1994). Similarly, industrial furnaces and boilers, especially those that burn hazardous wastes, could also generate airborne dioxin emissions (USEPA, 1985). The EPA made no estimate of emissions from these sources.

- *Cigarette smoke.* Dioxin has been found in cigarette smoke (Muto, 1989). When a cigarette is burned, several factors contribute to dioxin formation in the smoke. First, chemicals are added to cigarettes for a variety of reasons, such as to promote "smooth even burning" or to act as preservatives. At Congressional hearings held in March 1994, hundreds of chemicals were identified as present in cigarettes including at least thirteen ingredients that "are not allowed in foods that Americans eat" (Raloff, 1994). Some of these chemicals contain chlorine. Second, the tobacco is often sprayed with pesticides that contain chlorine. Third, the paper is contaminated with dioxin because of chlorine bleaching. Very little research has been done to determine levels of dioxin exposure to the smoker, but it seems likely that these levels could be significant.

Contributors

Beverly Paigen, Ph.D
Stephen Lester, M.S.
Charlotte Brody

Chapter Fifteen

Direct Organizing Strategies

If you're not fighting for what you want, you don't want it enough.

—Dave Beckwith, Field Consultant
Center for Community Change

Now that you've done the dioxin detecting work outlined in the last chapter, you're ready to start fighting for what you want. In this chapter we provide information on a variety of action campaigns to directly stop dioxin exposure, like shutting down incinerators and convincing pulp and paper mills to stop using chlorine. In the next chapter you'll find information on indirect campaigns to stop dioxin exposure, like getting involved in electoral campaigns and consumer efforts to buy only dioxin-free products.

At CCHW we believe that stopping the sources directly is most important, and that indirect efforts like buying dioxin-free products work best in conjunction with direct campaigns aimed at the major sources of dioxin exposure. We don't want our shared concern about the damage dioxin is causing to turn into individual efforts to create dioxin-free lifestyles. Stopping dioxin exposure should not be about learning to live without ice cream and hamburgers. It should be about learning how to build a society in which ice cream and hamburgers are free of dioxin.

Strategies to Stop All Incineration

Hundreds of grassroots organizations have worked or are working to prohibit any more incinerators from being built, to prevent any new kilns or boilers being permitted to burn hazardous waste, to stop waste fuels from being used in cement kilns, or to shut down existing incinerators. In general, it is easier to stop a bad idea from being put into action in the first place than it is to remove a bad idea that has already been put into action. But community efforts have been successful at doing both.

According to Ellen and Paul Connett's *Waste Not*, 280 proposals for medical waste incinerators have been defeated or abandoned since the 1980s and only 73 have been built—nearly a four to one ratio. Since 1993, companies have scrapped nearly twenty incinerator proposals in various stages of permitting and planning. These include proposals for trash burners, medical waste incinerators, and hazardous waste incinerators.

The incinerator industry has often claimed that unfavorable "market forces" have led them to conclude that building an incinerator is a bad idea. In all cases, they would have built their bad idea if not for the efforts of grassroots activists who have organized to stop the proposals. The ammunition provided by the EPA's report on dioxin makes stopping incineration more urgent and more possible.

Public education campaigns on the incinerator-dioxin link can help stop incineration. Build the argument against incinerators with data, outlined in Chapter Fourteen, that shows the primary role incinerators play in producing dioxin. A credible case against incineration can be built by describing how incinerators generate dioxin, identifying the worst incinerators, and showing how recycling is not only a safer but also a financially sounder alternative. According to a 1991 study by the Massachusetts Department of Environmental Protection, paper recycling added $518 million to the state's economy, creating nearly 9,000 new jobs.

Part of your argument against incinerators should include the huge financial costs associated with incinerators, which are often funded through taxpayer-financed municipal bonds. The failure of these bonds can hurt a city's credit, thus jeopardizing other economic development plans. The Detroit, Michigan incinerator cost the city so much that the facility was sold to Philip Morris at a disastrous loss to get rid of the financial burden. In many communities with large regional incinerators, the local residents pay more per ton for waste disposal than do the outside communities that use the incinerator. In Montgomery County, Maryland, trash collection fees of $146 per household will be raised to $246 by 1999 to support their $325 million incinerator. And, in Claremont, New Hampshire, local trash costs are $96.50 a ton, while outsiders can dispose of their trash at the same incinerator for $40 a ton.

There are several tactics you can use to fight proposals for incinerators.

- *Follow the dollars.* Find out who the financial backer is for the proposed or existing facility. If it is a bank, then you can ask the bank to not back the proposal or the facility. If the bank refuses, you can target it directly. For example, everyone can go to the bank on a busy day and withdraw money, or tie up the lines with people depositing one dollar at a time or opening new accounts. You can march in front of the bank, handing out flyers and talking about how the bank's support of the facility will hurt the local economy and the investors in the bank, given the number of bankruptcy and liability claims from such facilities. If it is a state bond, you can follow the example of New York City activists who got state legislators to redirect state incinerator subsidies towards recycling programs. This strategy helped to defeat the proposed Brooklyn Navy Yard incinerator.

- *Check up on the company's procedures.* If the company used local municipal bonds to support the incineration facility, check to see if it followed the proper procedures to obtain them. Local governments often don't pay much attention to their own rules. If the company hasn't followed proper procedures, the bonds can be stopped.

- *Ask about the waste facility's insurance policy.* If it is self-insured, it does not have much protection. Most insurance companies will no longer insure waste facilities for discharges and associated liability damages. Pose the question: If big insurance companies like Lloyd's of London won't insure the facility, can the plant really be safe? In Kentucky, residents demanded that millions of dollars in cash be provided upfront by the company as insurance before the facility began operations. The funds were to be managed by the local government. The company, unable or unwilling to put so much cash on the table, moved on.

- *Pass a local citizen referendum on the incinerator proposal.* In many communities, incinerator fighters have been able to turn their efforts into a citizen referendum. This helped educate the public about the group's concerns and provided a concrete way for people to express their beliefs through a vote. In several communities in New York, local groups put the question of a waste facility to a referendum vote and won.

- *Work for a statewide ban on incineration.* In 1992, Rhode Island became the first state to ban incineration as a solid waste disposal option. The province of Ontario, Canada, also enacted an incinerator ban around the same time. Several incineration moratorium bills have also been introduced in the U.S. House of Representatives. The 1994 version had over ninety co-signers.

- *Pressure your local government.* Incinerators can be stopped by the actions of local governments, pressured by organized residents. In the Chicago area, the Harvey City Council declared the town an incinerator-free zone in 1994 as part of the citizen effort to stop a 24-tons-per-day medical waste incinerator in the community. At the urging of the local community group, the New Bedford, Massachusetts City Council took bold action to prevent an EPA incinerator from operating within its borders. (See the sidebar in Chapter Twelve.)

- *Use local police powers creatively.* In Chickasaw, Alabama, the residents used local police powers to pass an ordinance governing the transportation of wastes to a proposed incinerator ship. The law dictated that trucks could not move wastes during certain hours because of children traveling to and from school, and that all trucks had to pass a safety inspection before moving through the town. The trucking companies were expected to pay the salaries of the inspectors. Each truck was required to have a police escort to travel through the town. This, too, had an associated cost to the trucking companies. The company decided not to dock their incinerator ship near Chickasaw.

- *Pass a "bad boy" ordinance.* If the incinerator company has a record of felony convictions, then they can be deterred by passage of a "bad boy" law. This type of law says if a company has been found guilty of a felony, it cannot do business in your town, city, or state. (For an example, see Appendix C.) This has kept polluters out of many communities.

- *Show your community its future.* Line up trucks or circle trucks through the neighborhood to demonstrate the noise, traffic, and disruption that a waste facility entails. This is especially powerful if the proposed truck route is in the same area where children will be waiting for school buses or walking to school. Ask the local Parent-Teacher Association to help your group stop the incinerator and its poisonous discharges, as well as its added traffic risks to children.

What invariably defeats incinerator proposals is a vigilant citizenry which organizes, raises public awareness, and exerts political pressure. These proposals are typically brought to a community in very hushed and very quick-moving discussions between the corporation and local government. The only thing that prevents these fast-track efforts from panning out is citizens who ask questions, raise issues, and organize.

Closing down an existing facility can prove to be more problematic than defeating a proposal for a new facility. And even if you succeed, closed incinerators have a way of attracting other toxic proposals for their idle smokestacks.

In a few instances, incinerators will almost close themselves down, as they rack up endless records of violations, fines, upsets, and explosions. But again, when a watchdog citizens group continually exposes these incidents and keeps public pressure on regulators, there is a much greater chance that the facility will close.

Faced with an EPA order to test for dioxin from their municipal waste incinerator, city officials in Akron, Ohio, voted to shut down the plant rather than face the likelihood of documented dioxin emissions at the facility. In Dayton, Ohio, officials were told to either retrofit or close down their two trash burners. They chose to close the burner that had tested high for dioxin in 1989. The other incinerator will be upgraded with new pollution control devices in an attempt to meet new EPA standards, and will remain open.

Strategies to Stop Medical Waste Incineration

Back when the world was just beginning to understand that germs cause illnesses, the medical community was surprised to find out that hospitals were a major source of contamination—as medical personnel spread contagious diseases from patient to patient on their unwashed hands and dirty equipment. Hand-washing and sterilization became accepted as necessary practices to prevent the spread of diseases.

Now that dioxin has been identified as a major chemical pollutant that is harming our health, hospitals have once again been identified as a major source of health problems.

The American Hospital Association has responded to the EPA dioxin reassessment by taking "strong issue with both the scientific method used to conduct the reassessment and with the preliminary conclusions drawn by the EPA" (AHA 1995a). Rather than deny that a problem exists, hospitals should be developing a preventive solution to the dioxin health emergency.

The EPA's new medical waste incinerator (MWI) regulations are touted by the agency as the way to clean up this largest source of dioxin. But these new standards would still allow the release of 110 grams TEQ per year of dioxin and furan into the air, not including emissions from new commercial regional medical waste incinerators. The rules do not account for the increase in dioxin in ash and effluents that will result from the more sophisticated pollution control equipment required by the new standards. And the EPA's new standards will also allow the smaller MWIs that release disproportionally more dioxin to continue their unsafe, substandard operations (PSR, 1995).

The EPA has estimated that its new regulations will more than double the cost of medical waste disposal, from $168 to $390 per ton. These estimates are based on the assumption that the new regulations will cause hospitals to shut down their on-site incinerators and ship their waste to much larger, regional, "state-of-the-art" incineration facilities run by companies like BFI and Waste Management (now WMX). These regional facilities are often located in low-income communities and communities of color.

As hospitals consider the relative costs and benefits of incinerating on or off site, your Stop Dioxin Exposure Campaign coalition can provide them with a third solution to their waste problem:

- An immediate halt to incineration of anything but pathological (human tissue) waste.

- Implementation of alternative disinfection methods (autoclaving and microwaving) for infectious wastes.

- Substitution of other materials for polyvinyl chloride (PVC).

- Establishment of safe materials policies which substitute reusable items for disposables and maximize the reduction, reuse, and recycling of non-infectious waste.

The U.S. Congressional Office of Technology Assessment (OTA) reports that autoclaving (steam sterilization) and microwaving are fully adequate disinfection technologies, and that subsequent shredding can reduce waste volume by 60 to 80 percent. The OTA also found that these treatment methods significantly reduce costs (USOTA, 1990). The only wastes that are not suitable for treatment by these methods are pathological wastes (body parts and other tissue-derived wastes), which may be burned without significant pollution hazards.

The majority of the chlorine in medical waste, which produces dioxin when incinerated, comes from polyvinyl chloride (PVC) plastic in packaging, gloves, infusion bags and tubing, bedpans, trays, ID bracelets, gloves, and numerous other medical products. Plastics, mostly in the form of PVC, account for 14 to 30 percent of non-infectious medical waste (Hickman, 1989).

Eliminating PVC is critical to preventing dioxin emissions from MWIs. Strategies should focus on substituting alternative materials for PVC in hospital supplies as part of a greater hospital waste management and reduction strategy. Materials management and purchasing departments will need to specify PVC-free preferences to their suppliers, as well as negotiate with manufacturers to eliminate PVC in their products.

In some instances, acceptable alternatives to PVC may not be readily available because suppliers are reluctant to make the switch. Public health professionals may be allies in convincing medical supply manufacturers (and trade associations such as the Health Industrial Manufacturing Association, which represents manufacturers of medical devices) to produce safer alternatives.

Suppliers who provide PVC alternatives should be acknowledged and encouraged. McGaw Inc. uses polyethylene to manufacture IV bags; W.R. Grace has developed a polyolefin blood bag. But Abbott Laboratories continues to use PVC to make IV bags and related supplies and has supported IV bag recycling efforts. Baxter has taken a similar approach. PVC recycling is an ineffective strategy because it encourages the medical facility to keep using PVC, which inevitably finds its way into the waste stream.

Many hospitals have already begun to reduce their waste by substituting reusable materials for disposable products such as linens, towels, gowns, draping, bedpans, scissors, food service equipment (cutlery and containers), dressing trays, surgical packs, and solution/water bottles. This is being done without compromising sanitary concerns (Mueller, 1994; Tieszen, 1992; USOTA, 1990; French, 1994; Belkin, 1993; Riggle, 1994). One study reported that substitution of reusable products for disposable and recycling of paper would result in a 93 percent reduction in surgical waste volume (Teiszen, 1992). Separation programs can be instituted to keep non-infectious waste apart from infectious waste, to help maximize recycling of disposable items that are not replaced with reusables.

Substituting reusable for disposable products is reported to lower costs significantly, since the cost of washing or disinfecting items for reuse is less than the cost of purchasing new items (DiGiacomo, 1992; Mueller, 1994; Tieszen, 1992; USOTA, 1990). One analysis found that a single teaching hospital saved over $100,000 per year by returning to reusable scrub suits and gowns in the operating room (DiGiacomo, 1992).

To start a campaign to end dioxin releases from MWIs:

- *Define your goal(s) before you begin.* Do you want to stop the incinerator? Do you want to make sure the hospital uses alternative treatment methods, starts a waste reduction program, or goes all the way to a PVC-free procurement policy?

- *Contact the hospital.* Find out what it does with its waste. If the hospital has an incinerator, is it looking into alternatives? Does it have a waste reduction program? Does the purchasing department have a policy to eliminate PVC where possible?

- *Talk to the staff.* In many hospitals, nurses may be good allies to help form waste-reduction or recycling committees. Nurses know how much waste exists in hospitals, how much plastic is used, and who to talk to about these issues. Find out if there is a nurses' union at the hospital that you can contact to help identify people who might be interested in working with you. Also try to identify and talk with staff in housekeeping, supply procurement, infection control, administration, and regulatory compliance. These hospital staff people may have a union as well.

- *Talk to the hospital's board.* Begin by contacting the hospital's board members who have visibility in the community and

Struggle for Survival in the South Bronx

The Bronx Clean Air Coalition in New York City was born in the fight against a medical waste incinerator that was built in our community without our knowledge. Starting in the summer of 1991, mothers, grandmothers, youth, clergy, civil rights advocates, and environmental justice activists from our community joined forces with the Riverdale Clean Air Committee, the Mosholu Woodlawn Community Improvement Association, and the National Congress for Puerto Rican Rights. These major groups formed the Bronx Clean Air Coalition, the first grassroots environmental justice coalition in New York City. Membership includes over sixty community-based organizations and hundreds of individuals. Together, we have been turning the heads of New York City's politicians and regulators nonstop.

All of us have become environmental justice experts in what seems like a flash of lightning. We had to. Through our "Saturday Outreach" program, rallies, health fairs, voter registration drives, and numerous

power to affect hospital policy. Many hospitals are also part of a regional hospital system, which may be a good first point of contact in attempting to get institution-wide policies going.

- *Use hospital competition to your advantage.* In most towns, local hospitals are competing with each other for the trust of the community. If you see hospital advertising on television and in the newspaper, competition is a factor in your community. "Hospital wars" mean that the hospital will be concerned with its public image as a community polluter. They also mean open houses, nutrition classes, and lots of other public events at which your coalition can raise the incinerator issue.

- *Talk to other activists.* National organizations such as CCHW and Greenpeace can put you in touch with other activists working to stop medical incineration. Physicians for Social Responsibility is working on toxic chemical issues: and this group's local members may also be useful contacts.

Strategies to Stop Chlorine Use and Production

In 1992, the U.S.-Canadian international Joint Commission on the Great Lakes concluded that "the use of chlorine and its compounds should be avoided in the manufacturing process" (IJC, 1994). In 1993, the American Public Health Association called for a phaseout of all organochlorines (APHA, 1993). Greenpeace has called for governments to plan the phaseout of chlorine, with priority given to the largest and most dangerous uses for which alternatives already exist: PVC plastic, pulp and paper, pesticides, and solvents. It has also called for an end to the burning of chlorinated compounds in incinerators. In the absence of strong federal government action, your local Stop Dioxin Exposure Campaign can tar-

marches, the Bronx Clean Air Coalition has been able to educate hundreds of people on the environmental concerns of the community. These efforts have paid off, as the Bronx Clean Air Coalition and its supporters represent an incredibly diverse network of concerned citizens.

The South Bronx is at times referred to as the poorest congressional district in the United States. Our community has one of the highest rates of infant mortality, compromised immune systems, lead poisoning, and asthma. This comes as no surprise, because the South Bronx is littered with numerous polluting facilities.

In our immediate area, there are over sixty-five waste transfer stations transporting asbestos, lead piping, construction debris, medical waste, sludge, and other toxins from all over the United States.

And there is the medical waste incinerator—originally a creation of Resource Management Technologies (REMTECH), a subsidiary of Montenay, and sponsored by Bronx Lebanon Hospital, a local conglomerate. Now REMTECH is bankrupt, and waste giant Browning-Ferris

Industries is poised to become the next owner of the facility.

Since the incinerator's inception, the New York State Department of Environmental Conservation (DEC) has deliberately sidestepped the community's concerns, operating in a typical closed loop fashion, in which critical evaluations and decisions are made by politicians and crooks who have used the guise of "economic development" to sell death to our community. The coalition has opposed the facility on the grounds of fraud and damage to the environment.

We not only continue the fight to close the medical waste incinerator, but also actively work to stop the pollution in our community and bring in sustainable development and affordable housing. With groups such as the Voter Participation Projects of the Community Service Society, the Northeast/Puerto Rico Environmental Justice Network, the Network for a Sustainable New York, CAFE, and the New York City Environmental Justice Alliance, we have brought the issue of environmental racism and injustice to the forefront in this city. It is up to us to save ourselves and our families.

—*Nina Laboy*

get individual producers and large-scale users of chlorine.

The pulp and paper industry is the second largest consumer of chlorine. Safe and effective alternatives are readily available to this highly polluting industry. Totally chlorine-free (TCF) bleaching is now effective, feasible, and economical. Because the technology is already readily available, eliminating dioxin from pulp and paper mills means changing processes, not closing mills or eliminating jobs. Conversion requires a capital investment, but a mill can offset that cost in just a few years through reduced expenses for energy, wastewater treatment, sludge disposal, liability, and remediation (Greenpeace International, 1995).

Environmental groups have been pressuring the industry to eliminate chlorine use in paper production for years. And the U.S. pulp and paper industry has already taken some steps to phase out chlorine bleaching. It claims to have already voluntarily cut chlorine use by about 70 percent. Additional pressure can make the industry become chlorine-free, and encourage it to adopt closed loop systems that reap further savings in water and chemical use. Halfway efforts like using chlorine dioxide (elemental chlorine free, or ECF) still produce dioxin (but in smaller quantities) and other toxics, and don't allow for closed loop systems.

PVC plastic production entails the largest single use of chlorine in the United States. It accounts for about 30 percent of all chlorine produced, or almost 4 million tons. Viable alternatives exist for nearly all of the products made from PVC. Manufacturers and users need to be pressured to convert to these alternatives.

Procurement and consumer campaigns to decrease demand for PVC will be discussed in the next chapter. Direct strategies to stop chlorine use in the production of PVC, pesticides, and other chemicals include getting local producers and users to agree to phase out the use of chlorine or to stop its use by a certain date. A good neighbor agreement between the community and the plant can be the mechanism for this chlorine phase-out or ban. A sample good neighbor agreement is in Appendix E. Another way to focus pressure on chlorine production and consumption is to implement a surcharge on chlorine manufacture and use. Chemical manufacturers already pay such a charge on toxic waste, based on the amount of chemicals they produce, which pays for the Superfund toxic waste cleanup program. A similar charge on chlorine could go to a Superfund to assist workers in the industry transition into similar-paying jobs outside the chlorine industry.

Another tactic to stop chlorine use and production is to press for local or state laws that prevent chlorine makers or industrial users from receiving government subsidies. Most major industrial facilities receive some type of governmental tax dollar giveaway, either in tax breaks, utility subsidies, or direct funds. Alternatively, the continuation of a subsidy could be based on an agreement to phase out or ban chlorine use by a certain date.

Many of the strategies to end incineration can be applied to chlorine use and production, including closely monitoring the regulatory process, putting pressure on the financial backers, and enacting bad boy ordinances and/or strict local rules on emissions.

Strategies to Secure Environmental Justice

The elimination of dioxin is an environmental justice issue. Cases of assaults on low-income and people of color communities must be exposed and challenged. A series of studies have documented the disproportionate number of polluting facilities, toxic waste sites, and chemical emissions in poor communities and communities of color. The most recent of these was released on August 25, 1994. The study, sponsored jointly by the Center for Policy Alternatives, the United Church of Christ, and the National Association for the Advancement of Colored People, looked at 1993 census figures and found that, as with 1980 census data, the percentage of people of color is three times higher in areas with the highest concentration of hazardous waste facilities than in areas without a commercial hazardous waste site (Goldman, 1994).

Robert Bullard, author of several books on race and the environment and Director of the Environmental Justice Resource Center at Clark Atlanta University, defines environmental racism as "any policy, practice, or directive that, intentionally or unintentionally, differentially impacts or disadvantages individuals, groups, or communities based on race or color." (See Appendix D for "The Principles of Environmental Justice.")

Beginning with the First People of Color Environmental Leadership Summit in 1991, groups have used different strategies to counter environmental racism, from challenging the racism inherent in local siting and cleanup struggles all the way to securing executive orders from the White House.

Notorious polluter WMX Technologies tried to jump on the environmental justice bandwagon. WMX scheduled a symposium on Environmental Justice for June 1994 at their Chicago-area headquarters. They falsely listed several leaders from the environmental justice movement as speakers. But grassroots groups forced a cancellation of their cynical effort.

Grassroots groups in the Chicago area organized to expose this sham event. Residents in this area, many of whom are low-income people and people of color, live with countless sources of industrial pollution and dumps of all kinds, and are fighting at least seven incinerator proposals. A prominent dumper here is WMX. Led by the People for Community Recovery's Hazel Johnson, activists shamed WMX into canceling the symposium.

Environmental racism became a major factor in Kettleman City, California, an overwhelmingly Latino community where WMX-owned Chem Waste Management runs a hazardous waste dump. For six years, until late 1993, WMX planned to site a hazardous waste incinerator in the community. Among the many strategies used by El Pueblo Para El Aire Y Aqua Limpia (People for Clean Air and Water) was invocation of Title VI of the U.S. Civil Rights Act. The group filed a suit claiming the toxic presence of WMX in their community was a violation of their civil rights. WMX backed out of the incinerator plan before the suit could proceed.

Now Kettleman City organizers have joined with activists from Westmorland and Buttonwillow in challenging environmental racism. These majority Latino towns are the sites of the only three hazardous waste dumps in the state. Groups from the three communities have filed suit under the 1964 Civil Rights Act to stop the siting and expansion of any toxic waste facility in communities of color. Targets of the suit are the state and local governments, Chem Waste Management, and Laidlaw, which operates the facilities in Buttonwillow and Westmorland.

For the first time in Texas, environmental justice issues were considered in the permitting proceedings for a proposed polluting facility. As a result of pressure from grassroots activists, several days of testimony were heard on allegations of environmental racism in hearings on a proposal by WMX to expand their Skyline dump in Ferris, Texas. WMX is seeking to expand this dump right into an African-American neighborhood in the small town outside of Dallas. The Texas Natural Resources Conservation Commission (TNRCC) was to consider whether racism was part of WMX's plan

Pensacola, Florida

Rosewood Terrace, Oak Park, Goulding, and Escambia Arms are African-American residential neighborhoods in Pensacola, Florida. Escambia Treating Company (ETC) occupied twenty-six acres in the center of this area, separated from nearby homes only by a fence. ETC operated from 1943 to 1982, treating utility poles and foundation pilings with creosote and pentachlorophenol, and leaving behind highly contaminated soil, sludge, and groundwater.

For years, after heavy rains the ETC waste ponds overflowed into the streets and nearby yards. In 1991, nine years after the facility closed, the U.S. EPA appeared on the scene. Under emergency removal authority, the EPA excavated the contaminated area and stockpiled the wastes on-site. This was the birth of what local residents now call Mt. Dioxin, a thirty-foot-high, plastic-covered mound of the dioxin-laced waste.

Margaret Williams was outraged at what the EPA did in her neighborhood. She said, "The reckless nature of EPA's activity was almost as frightening as the contaminants themselves. No measures were taken to protect nearby neighborhoods. Choking fumes and poi-

for expansion, and if the company's widely disparate offers for residents' homes that sit adjacent to the dumpsite were further evidence of racism. In early 1995, the TNRCC made its decision, choosing to ignore the evidence of environmental injustice and grant the permit expansion. Meanwhile, an FBI investigation into allegations of bribery and corruption by WMX continues.

In Chicago, years after the University of Chicago Hospital was blocked by local activists from building a medical waste incinerator, the hospital contracted with a waste disposal company that has targeted Harvey, a poor African-American suburb, for construction of a commercial MWI. Citizens in Harvey, led by HACO, a local community organization, have so far been successful in stopping the incinerator. By challenging the environmental racism that led to the incinerator siting proposal, HACO has gained the support of the local city council.

Putting a name on environmental racism is speaking truth to power. That truth allows you to build coalitions with other social justice groups and puts corporations and government on the defensive.

Strategies to Include Workers

Any campaign directed at stopping dioxin exposure must consider potentially impacted workers. Roundtable participants recommended that a high priority for the campaign be the creation of provisions for workers who may lose jobs due to shifts in industrial production as dioxin sources are systematically eliminated.

Since the late 1980s, corporations have been increasingly successful in polarizing environmental and labor interests. Dioxin exposure cannot be eliminated if these divisions, drawn by corporate spindoctors, keep workers and community activists apart.

sonous dust have caused respiratory distress, persistent skin rashes, burning eye irritation, and made existing health problems even worse." Even the EPA recognized the danger at the site. An internal EPA memo stated, "The ETC Site in its present state continues to pose an immediate risk to public health..."

In March 1992, residents in the surrounding community began to meet, and organized Citizens Against Toxic Exposure (CATE). Through their collective actions they finally had the site changed from "emergency authority" status and placed on the "real" Superfund list. This gave the residents a right to participate in cleanup decisions at the site. For now, the EPA has decided to incinerate the waste on-site.

CATE members have been working to obtain relocation for the residents trapped in this nightmare. CATE members believe that if they were a white community they would have been moved already. The dioxin level used at many sites to determine whether relocation and emergency protective public health actions should be taken is 1 part per billion.

Throughout this community, the levels of dioxin far exceed the 1 ppb trigger level, measuring "100 to 1,000 fold over the action levels that were established to prevent an immediate threat."

CATE members have placed little white crosses in the front yards of families who have lost a loved one to cancer or other dioxin-related diseases. Residents point out that Mt. Dioxin is literally across the fence from their backyards. According to CATE, twenty-eight people have died in the community since the cleanup began.

Workers in dioxin-producing facilities are not the enemy, nor should they be the target of our campaigns. The decision to continue or stop dioxin exposure is made in the corporate boardroom, not on the factory floor. Workers can be potent allies in the struggle to reach these decisionmakers and eliminate dioxin exposure.

Actions undertaken by a community/worker coalition focused on workplace conditions can be an integral component in your campaign to stop dioxin exposure. After all, workers in a plant are closest to the production processes, and therefore are exposed sooner and at higher levels than those in the community. The following strategies may help you build community/worker solidarity:

- Talk with the workers and/or the union about dioxin and try to find some common ground. Are the workers getting sick? Would the workers like to have a workplace study conducted to determine whether they are getting sick? How can your group help them obtain such a study?

- Investigate the finances of the plant and the corporation. If new measures were taken to stop emissions, would it remain financially solvent? Is the plant's production diverse enough that chlorine manufacturing or use is a small component of its overall operation?

- Ask the workers how they feel about a good neighbor agreement with the plant. This agreement can include notification to the community of dioxin releases, community representation on an oversight committee, a promise of new equipment to reduce exposures to the community and workers, and a list of issues defined by workers (see Appendix E).

- Investigate subsidies. Chlorine producers in the Niagara Falls, New York, area receive large electricity subsidies. Occidental Chemical receives an equivalent of $19,733 per year per employee in energy subsidies. Olin received $25,894 per year per employee. Encouraging public scrutiny of tax dollar giveaways to corporations can be an effective way to generate support among taxpayers and workers. Community leaders can organize to have representatives sit on the public board that makes the decisions about subsidies. Or the community can raise issues with the existing board about a particular corporation that is receiving subsidies but not being a good neighbor or employer.

- Your coalition can influence the subsidy board to secure subsidies and research-and-development funds for non-dioxin–related industries. Or you can restrict subsidies going to dioxin-producing industries so that they can only be used for research and process changes to reduce dioxin formation and discharges. Workers will be important allies in defining, and then advocating for, safe alternatives to dioxin-producing practices.

- Join in efforts to reform Occupational Health and Safety Administration (OSHA) standards and gain more stringent implementation of existing OSHA rules to protect workers from dioxin and other workplace chemical exposures.

Unions, like religious organizations, sometimes have pension funds or other stock portfolios that include major holdings of certain corporate stocks. These holdings can then help you to implement shareholder strategies to influence corporations to reduce dioxin and keep jobs.

Labor-environment coalitions are emerging across the country. In Louisiana's Cancer Alley, Oil, Chemical and Atomic Workers (OCAW) Union members and community groups joined

Jay, Maine

In March, 1987, when workers at International Paper Company's mill in Mobile, Alabama rejected the company's contract proposal, the company locked out all 13,000 workers and replaced them with scabs.

Three months later the 1,250 members of the United Paperworkers International Union Local 14 at International Paper's Androscoggin paper mill in Jay, Maine, went on strike, in response to the lockout at Mobile and local concessions demanded by the company. Over the ensuing months, as IP permanently replaced the entire workforce and continued to threaten the health and safety of the community with massive chlorine and chlorine dioxide leaks, a real community of workers formed. They no longer accepted the idea that you had to trade the environment for good jobs. They decided to fight for both: jobs and the environment.

It didn't take the union long, with IP's unwitting help, to show that the company would sacrifice the environment to impose its will. The newspapers reported forces after OCAW was locked out of BASF's chemical plant in Geismar. In Pennsylvania, on the third anniversary of the Three Mile Island radiation release, over 15,000 trade unionists and their supporters marched in solidarity with local residents, demanding public health protection and jobs.

The New York State Labor and Environment Network, started in 1989, targets mutually beneficial projects and holds annual educational conferences. According to its mission statement, "Those who work for a living and those who live in communities endangered by reckless environmental practices are the same people. Unresponsive government and businesses that put profits before people are the common obstacles to the environmental and labor movements. Taking up each others' demands, working side-by-side in solidarity is the key to our success."

The effort has paid off in at least one previously divided community. Anne Rabe, of the Citizens Environmental Coalition, describes success in the tannery town of Gloversville.

One of the best examples of our success was once one of our worst failures. Years of bitter disagreement between union members in the Amalgamated Clothing and Textile Workers Union (ACTWU) and the local environmental group, Rainbow Alliance for a Clean Environment, centered on job blackmail, serious pollution, and worker hazards. We asked leaders to face the tensions and discuss what had polarized their community. At that time, Bill Towne of ACTWU noted that "There is at least one thing we have in common: management blames us both. Maybe we should use that as a starting point for talking to each other." The two groups then started talking, worked on a tannery odor problem together and now are allies. Towne said, "It's always been—is it jobs or the environment first?—and we've ma-

that on the first day of the strike, a clearly visible brown liquid flowed from IP's effluent pipe. Many times after this incident, IP would have a spill or a leak and the union would bring it to the public's attention, only to have state officials support IP's contention that nothing had happened.

Jay workers started to lobby the town's selectmen to get something done. As a result, the Town Select Board formally notified IP that the board was reversing its decision to proceed with a $4.5 million bond issue. Also, it announced that three ordinances would come before a special town meeting. These would ban the use of professional strikebreakers; prevent IP from housing people on site; and require town officials to ensure that the state enforces environmental laws and ordinances affecting citizens or businesses. The town meeting was over in 29 minutes. About 900 voters approved overwhelmingly the three ordinances; you could feel the breeze created when 900 people raised their hands at the same time.

The Jay Select Board, in response to community pressure, instructed the town attorney to come up with an ordinance that would give the town the power to enforce state and federal environmental laws. The Jay Environmental and Improvement Ordinance, enacted in May 1988, encompasses all state and federal environmental laws relating to land, air, and water pollution. The town assumed the responsibility to license, monitor, and enforce pollution. Workers in Jay now have a full-time Environmental Administrator. Their environmental "group" is their town government. Their advocate does not raise money through bake sales and dances, but through taxation. Best of all, IP pays over 80% of the taxes. And the town plays no favorites: it fined IP $390,000 in 1993 for violation of the company's air license.

—*Peter Kellman*

tured enough to where we recognize that it's both (Rabe, 1992)."

Strategies for Cleanup of Contaminated Sites

Dioxin contamination and other serious chemical contamination at sites around the country has proven to be a highly charged political issue. Polluters don't want to admit liability and pay the bill for cleaning up. Developers stay clear of the liability attached to abandoned toxic sites. Government drags its feet, adding to the inertia that keeps these contaminated areas forsaken sacrifice zones. Local residents live with the health problems from past exposure and the threat that cleanup will expose them even more.

"How clean is clean?" is a loaded question at contaminated sites. Often the residents living nearby aren't included along with agencies and polluters in answering this question. And with the EPA now making development of contaminated urban areas, known as "brownfields," a feature of its clean up approach, it is very possible that many sites will receive inadequate remediation in the rush to wash clean corporate liability and redevelop these sites.

A good starting list of requirements for contaminated site clean up would include: (1) making polluters accountable for their contamination by retaining joint and several liability (which means that all companies and individuals that polluted are responsible), and requiring polluters to pay for the treatment of their waste on-site; (2) compensating victims of toxic exposures; (3)

prohibiting incineration as a cleanup option; (4) providing relocation for affected residents; (5) making cleanup jobs available for
area residents; and (6) crucially, involving the community as an
equal player in the entire decisionmaking process.

The two main targets in achieving these goals are the corporations and the government. Here are some ideas for activities your
group can take to pressure these targets to give you what you want.

- If the facility is still operating and the workers are organized,
talk to the union. If there is no union, talk to groups of workers.
Ask them for information. Maybe they have been fighting to
clean up the workplace and could use your coalition's support.
Or maybe they are negotiating a contract agreement with the
company and the outside pressure could be useful. They may
know how the plant could be modified to eliminate future
dioxin releases or how to best clean up existing contamination.

- Talk with former workers, especially if the corporation is no
longer in business. Usually, when a plant closes the workers get
a bad deal. Workers who have been hurt by their company are
usually more willing to talk about what, where, and how the
company dumped waste.

- Hold a demonstration and press conference at corporate or
government headquarters. At Love Canal, residents crashed
the company's public open house and picnic. Occidental
Chemical (formerly Hooker Chemical) wanted to improve its
corporate image through the open house, with press coverage
of officials giving speeches about what a good corporate neighbor the company was. The public needed tickets to attend the
event. Over 100 Love Canal residents and supporters obtained
them. Everyone wore t-shirts that said, "Love Canal, another
product from Hooker Chemical." In the middle of the picnic a
Love Canal resident blew a whistle and everyone dropped
where they were, falling to the ground to symbolize death. The
press coverage of the open house featured the die-in rather than
the "good corporate citizen."

There are often many levels of government agencies involved in managing a contaminated site, each having some power

Chemical Weapons Working Group

The U.S. Army is proposing to spend over $11 billion taxpayer dollars to create the largest hazardous waste incineration program in history to burn its stockpile of 30,000 tons of chemical weapons. Citizens living near chemical weapons stockpile sites want nothing more than to have these deadly weapons destroyed immediately. However, the issue of how to dispose of these weapons remains a complex debate between military and government officials and concerned citizens. The Army has proposed to burn these weapons in incinerators in the communities where they are stored. Geographically, the storage sites are located predominantly in politically disadvantaged areas and communities of color. Stockpiles are stored in Maryland, Alabama, Oregon, Indiana, Arkansas, Colorado, Kentucky, and Utah. The U.S. Army has already begun burning chemical weapons in the South Pacific.

The decision to burn the most lethal chemicals on the planet was made without citizen input. Citizens in communities storing chemical weapons have expressed concern over this plan since 1984, when the Army over the containment and cleanup of the site. You need to consider each level and decide which elected official or agency can give you what you want.

- Some groups have put together alternative cleanup workshops. At such a workshop, the group invites cleanup companies that offer technologies that will contain and not spread the dioxin and other waste. Most vendors promoting such technologies will pay their own way to and from the workshop. This workshop provides information for your coalition on cleanup processes. Elected officials and agency decisionmakers in charge of defining the cleanup plans should be invited to these events. If they come, they've learned as well. If not, you can take the information to them. Alternative cleanup workshops were successful in turning around bad cleanup plans in Stringfellow, California, Pitman, New

Jersey, and Lowell, Massachusetts.

- Superfund law requires the EPA to include public participation in cleanup decisions. One way the EPA fulfills this requirement is by holding several meetings to discuss cleanup options. This is an opportunity to use group actions to impact decisions. For example, one group of residents filled the hearing room, each person holding a 12-by-14-inch sign that read "NO." Every time someone mentioned incineration, regardless of who it was, everyone quietly raised their signs.

- The Downwinders group in Illinois marched through the areas downwind of an incinerator. They marched on the day the next town downwind of the incinerator was having its town council meeting. They ended the march at Town Hall and asked the town council to support their efforts to stop the incineration.

first announced its intention to incinerate these weapons. The Kentucky grassroots group Common Ground founded the Kentucky Environmental Foundation (KEF) in 1990 to advocate their concerns in national and international arenas. Since that time, KEF has built an international coalition of citizens groups called the Chemical Weapons Working Group (CWWG), a united force pushing for safe disposal of chemical weapons.

The CWWG advocates "closed-loop" chemical weapons disposal technologies such as chemical and biological treatments, which do not allow toxic chemicals to be released into the environment. We are having a tremendous impact! After a decade of working with local community groups, national environmental organizations, and scientists in pressuring Congress for safe disposal, we feel we are close to defeating the Army's incineration program. A victory in this battle will be a victory for us all!

—Elizabeth Crowe
and Craig Williams

- In New Jersey, a group released balloons with attached postcards which asked people to send the postcard to or call an elected official to advocate for the defeat of the incinerator proposal. Over 150 calls were received from families living over twenty miles from where the balloons were released. The people from these distant communities had no idea the wind would carry the balloons—and dioxin— that far. (Balloons do have their own environmental problems, which you should consider before you decide to take this action.)

Strategies for Toxic Waste Reduction

Over two dozen states, as well as the U.S. Congress, have passed pollution prevention laws since 1987. Pollution prevention laws, also called toxics use reduction laws, aim to reduce the use of toxic chemicals and the generation of toxic waste so that fewer environmental and public health threats are created. The National Environmental Law Center and the Center for Policy Alternatives recently evaluated pollution prevention laws in ten states. According to their report, *An Ounce of Toxic Pollution Prevention*, these laws differ significantly. Some concentrate strictly on toxics use reduction and require toxics users to plan on reducing chemical use in their processes. Others look only at the waste end, focusing on reducing the generation of hazardous wastes and establishing programs to assist waste generators in reducing the volume and toxicity of hazardous waste (NELC, 1993).

According to the report, an ideal pollution prevention law should include changes in production processes and products that reduce *both* the use of toxic or hazardous chemicals *and* the production of these chemicals as waste products. The goal of such changes should be to reduce all hazards associated with toxics use, including workplace, community, and consumer exposure. Chlorine would certainly be among the most significant toxics whose use must be reduced, and dioxin one of the wastes consistently measured to determine if the law is working.

The report also recommends that all laws include the following:

- Quantitative reduction goals for reducing the use of toxics and the generation of associated wastes, for the facility as a whole and for each production process.

- Toxic use reduction plans that are kept at the facility. State agency personnel should have access to it when they desire.

- Annual reporting on facilities' progress toward reductions.

- Plan review by workers, community residents, and local emergency planning committees of facilities that affect them. These groups should also participate in preparing the plans.

- Training seminars on toxics use reduction concepts and technologies.

- Citizen empowerment to sue facilities or enforcement agencies for failing to take actions mandated by the law.

- Requirements for study and monitoring of the impacts of toxics use reduction efforts on workers and communities.

- Strong penalties for anyone reviewing a plan who divulges important trade secret information to outsiders.

- Technical assistance by the state for toxics users, with priority determined by the processes' toxicity, widespread use, or potential for improvement (NELC, 1993).

We need you to add to the strategies in this chapter. What has worked in your community? What has failed? Use the suggestions in this chapter as an outline for your own efforts. Then let CCHW know how you tailored the strategies for your local campaign. We'll do our best to answer your questions and to put you in touch with other community activists going through similar struggles.

Contributors

John Gayusky
Charlotte Brody
Elizabeth Crowe
Peter Kellman
Nina Laboy
Charlie Cray
Joe Thorton
Lin Kaatz Chary
Craig Williams
Alicia Culver

Chapter Sixteen

Indirect Organizing Strategies

An indirect but powerful way to fight dioxin exposure can be to reduce the consumption of products whose manufacture, use, or disposal release dioxin. Chlorine-based products should be a major focus of indirect, consumer-end strategies. To phase out chlorine production, coalitions will have to publicize the link between chlorine production and chlorine-based products and dioxin exposure. The dioxin-chlorine link exists in scores of consumer products, including all PVC products, bleached paper, dry cleaning, pesticides, diapers, and tampons .

A targeted consumer boycott of dioxin-related products, and an aggressive push for chlorine- and dioxin-free alternatives such as "green cleaners" and totally chlorine-free paper, can put a dent in the chlorine demand that currently keeps plants running at capacity.

Another effective way to drive down demand for chlorine and chlorine-based products is through government procurement policies. Since governments are prodigious purchasers of paper and other supplies, its "demands" on the market will significantly impact the supply made available by producers. Stop Dioxin Exposure Campaign coalitions can develop local campaigns to

Totally Chlorine Free (TCF), Elemental Chlorine Free (ECF), Processed Chlorine Free (PCF)

Some pulp makers have responded to concerns about dioxin by switching from chlorine to chloride dioxide for bleaching. When chloride dioxide has been used, the pulp or paper is labeled ECF, or elemental chlorine-free.

While chloride dioxide produces less dioxins and other organochlorines than chlorine, the pollution it does produce is still extensive. Totally chlorine-free (TCF) paper is manufactured completely without chlorine.

Processed chlorine free (PCF) paper is recycled paper manufactured completely without chlorine. However, since the recycled stock used to make PCF paper includes paper that was chlorine bleached in its first life, the recycled product is never completely chlorine free.

force government and other large institutional buyers to purchase non-chlorine—bleached paper and non-PVC building materials. Your coalition can also halt the spraying of chlorinated pesticides in schools and public buildings, and put pressure on retailers to stop using chlorinated chemicals.

Dioxin-Free Procurement Campaigns

Collectively, purchases by federal, state, and local governments equal nearly one-fifth of the nation's gross national product. The federal government alone spends about $200 billion annually on goods and services, and operates over 500,000 buildings. Making government procurement dioxin-free will significantly reduce the consumption of products that increase dioxin exposure.

A major focus of the dioxin-free procurement campaign can be unbleached, recycled paper products. On October 20, 1993, President Clinton issued an executive order to direct all federal agencies to purchase environmentally preferable products. Unfortunately, provisions in the order encouraging totally chlorine-free (TCF) paper products were defeated by paper industry lobbyists.

Only a few agencies, including the EPA, are voluntarily buying chlorine-free paper.

Ideally, paper products should contain the maximum possible content of post-consumer recycled fiber. There should be no chlorine used in converting waste paper to the recycled pulp. Another environmentally-sound paper is made from kenaf, a fast-growing plant that yields three to five times as much pulp per acre as trees, according to the U.S. Department of Agriculture. Straw, bamboo, and hemp can also be used to make paper.

The Center for the Biology of Natural Systems (CBNS), offers activists the following strategies for convincing local governments to buy chlorine-free, recycled paper products:

- Identify the local or state government agency responsible for making paper purchases, and the individual in charge.

- Request a summary of recent paper purchases, broken down to describe paper type, recycled content, and low-chlorine and chlorine-free products.

- Request copies of laws or regulations that specify purchase of recycled low-chlorine or chlorine-free paper.

- Determine how compliance with these regulations is assured. Request copies of compliance reports.

- Examine the regulations to see how they can be amended to maximize the procurement of recycled and chlorine-free paper. This usually means amending product specifications that either call for virgin fiber paper or require high brightness standards, which demand chlorine use in paper manufacture.

- Develop a strategy for improving the procurement program. This should begin with a campaign of public education about the importance of a procurement program that reduces dioxin exposure, lessens the amount of paper trash, and improves the local economy. With public backing, lobby the relevant agencies and officials for an improved program. Offer to help them make the improvements, for example, by helping to draft model legislation and contacting prospective suppliers of recycled unbleached or non-chlorine–bleached paper products.

Several states, including Vermont, Oregon, Massachusetts, and Washington, are purchasing paper products that do not create dioxin. For example, the state of Washington buys peroxide-bleached copy paper and envelopes as well as unbleached paper towels and napkins.

Several U.S. cities, including Seattle and Chicago, have followed suit. The Seattle City Council passed a procurement law in 1993 favoring non-chlorine–bleached paper products. It directs departments to "purchase recycled content photocopy paper that has not been bleached with a chlorine-based lightening process, including elemental chlorine gas, chlorine dioxide, or hypochlorite, when non-chlorinate bleached photocopy paper is readily available and similarly priced."

Other Strategies to Get the Chlorine Out of Paper

Since the early 1990s, Greenpeace has campaigned to convince the publishers of *Time* magazine to use chlorine-free paper for its weekly editions. Even though *Time* purchases so much paper that it literally sets the industry standard for paper producers, it has refused to use its mammoth purchasing power to require mills to deliver chlorine-free paper. To appease its critics, *Time* asked its mills to use chlorine dioxide in their bleaching processes, which Greenpeace labelled "the ultimate half-way measure."

Subscribers and other purchasers can join the demand that publications switch to chlorine-free paper. Consumers can also urge copy shops to use non-chlorine-bleached copy paper as their first choice. Kinko's Copy Shops is the only national chain to date that offers customers unbleached recycled paper (Hammermill Unity DP) at the same price as nonrecycled, bleached paper. Other copy shops should be encouraged to adopt similar policies. If copy shops refuse to use environmentally-responsible paper, you can bring in your own paper while your coalition continues to pressure them to offer it at a comparable price.

Consumer groups have begun to express concern about the health hazards of direct contact with products that are chlorine bleached—most notably coffee filters, tea bags, women's sanitary products, pizza boxes, paper plates, and diapers. Direct consumer pressure in Europe has caused several companies to market non-chlorine-bleached alternatives for these products.

Independent laboratory tests conducted in England found detectable levels of dioxin in common consumer brands of tampons. These brands contain fibers bleached with chlorine bleach. Chlorine-free tampons, panty shields, and pads (including Natracare products, which are made in Sweden) can generally be found in health food stores and environmental products stores. Consumers should demand that large U.S. manufacturers switch to chlorine-free bleaching technologies in order to make safe products more affordable to American women.

Strategies to Replace PVC and Other Chlorinated Plastics

Approximately 60 percent of the polyvinyl chloride (PVC) produced in the United States is used in construction. Millions of American homes, schools, and office buildings are built with vinyl siding, window frames, doors, flooring, wire sheathing, piping, and other PVC construction materials. When these buildings burn, toxic dioxin emissions are released.

Activists can campaign for government agencies to pass ordinances prohibiting the use of PVC in public buildings. Such ordinances would protect employees and firefighters from the hazards of toxic fires, and would create a level playing field for contractors who bid on construction projects. The precedent for such actions has already been set in Germany, where more than seventy local governments including those of Berlin, Munich, and Stuttgart, have phased out the use of polyvinyl chloride in public construction and government offices. One German hospital was built almost entirely PVC-free, with only a few exceptions where there were no suitable PVC substitutes.

The most common use of PVC in construction is in piping for water and gas and in sewer drainage. Alternatives to PVC pipes include cement, cast iron, and vitrified clay or polyethylene plastic for large outside drainpipes. Copper and stainless steel can be used for small indoor potable water lines.

Homeowners and municipal builders should use wiring that is covered with PVC-free sheathing and insulation. Some cities have already started to do this in response to dangerous subway fires. Toxic fires in the London Underground and the New York City subway system have inspired municipal governments to start purchasing PVC-free wiring, often made out of low density polyethylene (LDPE).

Another reason to avoid PVC-coated wiring is that it creates dioxin when it is "recycled." Copper wire is reclaimed through a smelting process in which the PVC sheathing is burned off and the copper is melted down. This process releases large amounts of dioxin.

It is not easy for consumers to identify the full range of products and office supplies that are made with chlorinated plastics. With the exception of bottles, most plastics used for packaging and consumer products are not labelled. Purchasers should ask suppliers to confirm that their products are PVC-free. An Austrian Greenpeace report noted that most of the large European office supply companies have already banned or are about to ban PVC from their product lines in response to consumer pressure.

Consumers should demand that corporations stop bottling products in PVC. Most bottles are labeled on their underside with the type of plastic that is used. PVC is labeled with a #3 or V, often accompanied by the chasing arrows recyclable symbol. (People are misled into believing that the bottles are recyclable in their community when they see this symbol. In fact, according to the American Plastics Institute, less than 2 percent of all PVC generated in 1993 was recycled.)

An informal survey of products found in a typical chain drugstore in Washington, D.C. found that many manufacturers unnecessarily use PVC packaging. Products packaged in PVC were shelved next to competing products packaged in chlorine-free plastic. In some cases, the bottles themselves were nearly indistinguishable.

Most shampoos and body care products are packaged in either PETE, #1, or high density polyethylene (HDPE),#2, two types of plastics which lack chlorine in their chemical structure. These bottles are also accepted in most municipal recycling programs. The presence of PVC plastic in the wastestream can contaminate PETE recycling. The use of PVC for packaging is completely unnecessary given the plethora of alternatives.

Community leaders should advocate for state and local laws to curtail the use of chlorinated plastics. A small number of communities in the United State have passed model ordinances prohibiting the use of PVC packaging.

Alternatively, products should be bottled in nonpetroleum-based "bio-plastics" which can be composted. In Germany, Wella is using petroleum-free biopol plastic to bottle its shampoo and conditioner. Products like these showcase an innovative, emerging packaging technology.

Strategies to Eliminate Other Dioxin-Producing Products and Practices

Solvents use 10 percent of the chlorine produced in the world. Industries and government agencies should adopt and implement pollution prevention programs designed to eliminate the use of chlorinated solvents. Large volumes of chlorinated solvents are typically used to clean engine parts and other equipment. Water-based or mechanical cleaning methods can be substituted for chlorinated solvents.

The largest single user of chemical solvents is the U.S. Department of Defense. Some federal facilities have taken steps to reduce their use and inventory of hazardous chemicals, generally in response to environmental contamination problems such as hazardous waste leaks and accidental chemical releases. For example, Tinker Air Force Base in Oklahoma has an ambitious pollution prevention program with a zero discharge goal—no hazardous chemical usage or emissions by the year 2000. The base's engine maintenance program has already set a precedent by phasing out all ozone-depleting chemicals and eliminating 300,000 pounds of hazardous chemicals.

Traditional dry cleaners use perchloroethylene, or "perc," which is a chlorinated solvent. Dry cleaners should be encouraged to stop using chemicals and start using steam-and water-based cleaning. Many items are labeled "dry clean only," but few of these fabrics will be damaged by other cleaning methods.

Some household paint strippers and removers contain solvents such as methylene chloride that are toxic to the nervous system. Water-based paint removers should be used instead.

Home builders, construction companies, lumber yards, and government facility managers should boycott the purchase of cement at kilns that burn hazardous waste. Nationally, cement kilns burning hazardous waste contribute 4 percent of the dioxin air emissions. Municipalities should follow the lead of the city of Fort Collins, Colorado, which does not allow city-funded projects to use cement manufactured in kilns that burn hazardous waste. This action was prompted by a citizens' campaign to prevent a local cement company from burning hazardous waste.

Builders and remodelers should also avoid using pentachlorophenol. "Penta" and other types of chlorinated wood treatment agents discharge dioxin into the environment when they are manufactured, used, and disposed.

Consumers, government facility managers, and businesses should also eliminate the use of organochlorine pesticides, which are routinely used for both indoor and outdoor pest control. While pesticides account for only a small portion of chlorine use, their application greatly harms the environment. PTAs should discourage schools from spraying chlorinated pesticides in classrooms. Workers should consider pesticide spraying a serious occupational health risk. Farmers should be encouraged to use organic farming methods.

Indirect strategies like dioxin-free purchasing campaigns can supplement our direct efforts to convince corporations to stop the dioxin poisoning. When industries discover that they can profit more from *not* polluting, they are much more likely to agree to our campaign's demands. For more information, see the *Dioxin-Free Products Guide* from Ralph Nader's Government Purchasing Project (GPP, 1995).

Dioxin in an Election Year

The dioxin issue provides an opportunity for honest debate about the overarching influence of corporations in public life, and about the need for elected officials to protect their constituents' rights to life, liberty, and clean food, air, land, and water.

In an election year, even people who never or rarely vote are more willing to talk and think about the role of government. Campaigns against dioxin exposure can be the framework for a larger conversation on how, as activists/writers Danny Cantor and Juliet Schorr put it, "we [can] get government off our backs and on our side."

One way to reach out in an election year is through the media—by conducting direct actions that get covered by the press; and by making your points on talk shows, radio call-in shows, and public access television, and in letters to the editor. Another way is to go directly to the people and talk to them about why they should be concerned about dioxin and how to vote their concerns. Every significant social movement from the Populists of the 1880s to the Wise Use Groups of the 1990s has grown through events or meetings at which a real live person talks about issues with potential recruits.

Those potential recruits make up a large proportion of the so-called swing voters. According to a *Times Mirror* September 1994 poll, titled *The New Political Landscape*, the voting public is divided into ten distinct political groups. Three Republican-oriented groups make up 36 percent of registered voters. Four Democrat-oriented groups add up to 34 percent. The largest block of swing voters, making up 19 percent of the electorate, are referred to as the New Economy Independents. This group is made up primarily of high school graduates who are underemployed and not optimistic, under the age of fifty, 60 percent female, and strongly environmentalist. According to this poll, candidates of either party need the New Economy Independents to win a majority (TMC, 1994). If stopping dioxin exposure can be made into a stated concern of these strongly environmentalist voters, no candidate can win without jumping onto our bandwagon.

How do you make dioxin an electoral issue? Identify a sound byte that describes your local dioxin battle: Shutting Down the Solite Hazardous Waste Incinerator. Getting the paper mill to begin conversion to totally chlorine-free, closed-loop production. Enacting a dioxin-free procurement policy. Make sure that your demands are clear and meaningful. This is a stop dioxin exposure campaign, not a regulatory rule dispute that puts 90 percent of the electorate to sleep in nine seconds flat. If your electoral efforts are to get all the candidates to agree to ask corporations to lower dioxin production, discharges, and disposal as they develop new technologies on their own timetable, what have you won? What will change? Who will be moved by the visionary truth you have spoken?

Be very, very visible. When someone tells you that you can't expect anyone to pay attention to dioxin during an election season, explain that election season is the most important season in which to talk about the health of our people and our society. Most electoral campaign events consist of the same speech at the same flag-draped podium. If your events are visually dramatic and easy to understand, you can be the long lead story on the evening news—the story that everyone, including the candidates, has to watch before seeing campaign coverage.

Include all political parties in your education work. Offer to do a presentation on dioxin at party meetings. Send your educational materials to party officials and everyone seeking public office. Make sure these materials have points that a candidate can use to explain why he or she supports your community coalition's position on dioxin. Invite candidates to come to a meeting of your coalition to discuss the issues, or ask candidates to come with you on a toxic tour of local dioxin polluters. Make sure that no one running for office in your community can honestly say he or she knows nothing about dioxin.

Make your specific campaign an issue in local races. Make sure your question is included in candidates' press conferences by getting local reporters interested or by having a member of your coalition ask the question as a reporter for the coalition's newsletter.

Have members of your group attend every public forum during election season wearing "stop dioxin" t-shirts or buttons and talking about the Stop Dioxin Exposure Campaign. Hold your own candidate's forum on dioxin. If you've done a good job of organizing your local campaign, you'll have too many people in the audience for the candidates to ignore.

Find out which candidates are getting money from dioxin-producing corporations, corporate leaders, and PACs. Publicize this information. Send it to the press and pass it out at candidate's forums and other electoral season events.

Know the law and don't break it. While election laws vary from state to state, there are some basic rules. First, there is a big difference between electoral and nonelectoral politics. A tax-exempt charitable organization under U.S. Internal Revenue Service 501(c)(3) rules cannot engage in electoral politics. A coalition made up of 501(c)(3) organizations cannot engage in electoral politics. The Internal Revenue Code states clearly that a 501(c)(3) organization can't "participate in, or intervene in a political campaign on behalf of any candidate." This means that the organization cannot endorse candidates, can't volunteer in a candidate's campaign as members of the Stop Dioxin Exposure Campaign coalition, and can't create slates of candidates or voters' guides that suggest which candidates are the good guys on your issues. But you can do all of the other activities described above to make dioxin an issue during election season. You can print the results of candidate questionnaires as long as you cover a wide range of issues, include all candidates, and don't state a preference for certain positions or certain candidates. You can also sponsor voter registration drives and get out the vote activities as long as they are done as non-partisan and neutral events. For more information, see Americans for the Environment's *Permissible Political Activities: A Basic Guide for Grassroots Environmental Activists Groups* (AE, 1994).

If you want to endorse or create a voters' guide only on the dioxin issue or sponsor drives to get people to vote for certain candidates, you need to register with the federal government as a 501(c)(4) or a political action committee (PAC), and in your state and county according to local election laws. In that case, you should make sure that all donors know that they can't deduct their donation to you from their taxes. Many non-profits have both 501(c)(3) and 501(c)(4) arms. NARAL, Planned Parenthood, Right-to-Life, NOW, and the Sierra Club all have a lot of experience in how to run two similar but different organizations at the same time.

If your group or coalition has not registered as a 501(c)(3) or a 501(c)(4), the only restrictions you have to worry about are those attached to state and sometimes local election laws.

Most eligible Americans are choosing not to vote for any elected officials. Making dioxin an electoral issue can also be a way to promote discussion in your community about how we came to be a country of non-voters. The EPA report and other recent studies provide an opportunity to talk together about why our current elected officials haven't taken more decisive actions to protect our health, our jobs, and our communities—and about what we can do to change this.

Contributors

Alicia Culver
Charlotte Brody
John Gayusky

Chapter Seventeen

What To Expect From Industry

We are up against an awesome case of money and power. It takes a long time when you have key people in government and business working against you.

—Admiral Elmo Zumwalt, Jr.

Admiral Elmo Zumwalt Jr. knows about the willingness of corporations and their friends in government to manipulate scientific studies and public opinion to hide the true risks of dioxin exposure. Zumwalt, who ordered the spraying of Agent Orange in Vietnam and whose exposed son consequently died of Hodgkins disease and lymphoma, has documented how the Office of Management and Budget (OMB), the Veterans Administration, the Centers for Disease Control, and the chemical corporations manipulated research in order to avoid finding negative health effects of Agent Orange.

Industry is doing its best to amass "an awesome case of money and power" to stop citizens from controlling dioxin exposure. In 1994, corporate contributions to the Chlorine Chemistry Council's (CCC) budget for public relations and research totaled

$12 million—up from $2 million in 1992. The CCC is the chlorine industry's lobbying and public relations coalition. The CCC has hired not one but two public relations firms to help shield dioxin polluters from the American people. Occidental Petroleum's president and CEO, J. Roger Hirl, promises that the chlorine industry could contribute "ten times" the current level of people and resources to the CCC if they need it to fight the campaign to stop dioxin exposure.

The CCC has been hard at work to prevent the EPA's dioxin report from ever becoming finalized. The council's efforts have centered on the Science Advisory Board (SAB). When the May 1995 meeting of the SAB's committee on the reassessment failed to produce any significant challenge to the findings of the reassessment, the CCC and its allies from the American Forests and Paper Association (AFPA) made up their own story of what happened at the panel's meeting and spun their story to the *Wall Street Journal* and other media outlets. These corporate maneuvers will likely delay the regulatory process on dioxin.

The AFPA and the CCC, along with the incinerator associations and their various public relations firms, can be expected to play major roles in trying to stop our Stop Dioxin Exposure Campaign. While the chlorine industry and incinerator advocates may occasionally blame each other for creating the dioxin problem, we should expect that, most of the time, their attacks on our Stop Dioxin Exposure Campaign will be the same. We can expect them to use six general strategies to counteract efforts to stop dioxin exposure:

- As long as we're not sure, we can't do anything.
- We know plenty about dioxin exposure, and there's nothing to worry about.
- Chlorine's benefits outweigh any dioxin risk.
- Personal responsibility, yes; corporate accountability, no.
- We'd like to stop making dioxin, but it is just too expensive.
- Poisoning is the cost of civilization.

Activists can work to counter each of these corporate excuses. Your coalition needs to educate your community on what the scientific studies do tell us. Then you need to inoculate the people you educate to prepare them for the corporation's strategies.

Medical inoculation, such as a polio or smallpox vaccination, is done by giving people a little bit of a germ or virus. The body has time to defend itself from such a controlled dose, by building antibodies and creating a resistance. So when an actual exposure to the germ or virus occurs, the body is prepared to successfully fight back. That's what your coalition needs to do with corporate arguments. Let your activists and your audience know what the other side is going to say. Give them a small dose of the opponent's argument. Then teach them what is wrong with that argument, and how to defend themselves against it. Remember, poll data show that people's instincts and experience will lead them to trust you, not the corporations.

What Is a Corporation?

Historically, in the United States, incorporation was regarded as a privilege—not a right—and corporations were understood to be legal fictions, subordinate to the public interest. For several generations after the nation's founding, both the law and popular culture reflected the common understanding that corporations were not chartered to cause harm, and if they did the penalty was extinction.

The people defined corporations and kept them in check by carefully writing the rules into corporate charters, state corporation laws, and state constitutions. For example:

- The law did not regard corporations as "persons."
- Corporations were chartered for fixed periods of time, such as twenty-five years. When charters expired, elected state legislators decided whether renewing the charters would be in the public interest.
- Corporate owners, managers, and directors were legally responsible—fully liable—for debts and harms their corporations

caused; some states made them doubly and triply liable.

- No corporation could own another corporation.
- There were limits on how much property a corporation could own, and on how much capital it could control.
- Corporations were prohibited from contributing money to electoral candidates.
- When corporations violated the public trust by exceeding their authority or by harming the common good, states revoked their charters, thereby ending those corporations' existence. After all liabilities were met, remaining assets were divided up among the shareholders.

By the early twentieth century, corporations had changed the law, and the circumstances of their own creation. Due to decisions

"As Long As We're Not Sure, We Can't Do Anything"

This is an elegant argument that allows corporations to assure the public that they will protect us just as soon as "sound science" indicating the need for such protection becomes available.

This strategy has long been a major part of the battles over dioxin and other environmental health issues. The 1994 EPA report was the result of the successful use of this strategy by industry in 1991. Now, four years and 2,400 pages later, industry consultants are still recommending that chlorine producers and industrial users protect themselves from evidence of dioxin toxicity by using the "gaps in science" to "build a scientific wall" to "highlight uncertainties."*

---

* This advice to the chlorine industry came from Dr. Terry Yosie, a consultant with the E. Bruce Harrison PR firm and a former EPA official and staff person at the American Petroleum Institute, at an April 1995 conference on "The Changing Chlorine Marketplace," put on by the industry journal *Chemical Week*.

In conjunction with the "as long as we're not sure" argument, corporate interests counter efforts to stop dioxin exposure by claiming that important environmental decisions, made before all the evidence was in, have resulted in costly and unnecessary public policy. Two people have been largely responsible for the implementation of this corporate strategy: journalist Keith Schneider and Vernon N. Houk, one-time assistant surgeon general for the Centers for Disease Control (CDC).

Keith Schneider has written numerous articles about dioxin for the *New York Times*, including a five-part March 1993 series on misguided federal environmental policies. Marc Smolonsky, the former chief investigator for a House Government Operations subcommittee that looked into the government's botched Agent Orange studies has described Schneider's coverage of the dioxin issue as "scandalous."

One Schneider article headlined "U.S. Officials Say Dangers of Dioxin Were Exaggerated" was reprinted by at least twenty major newspapers (Schneider, 1991). Other dailies

made by the United States Supreme Court in the late 1800s, corporations became "legal persons," gaining Bill of Rights protections. As a result, corporations today enjoy freedom of speech and other constitutional rights originally designed to enable free people to protect themselves from tyranny. Yet employees on corporate property do not enjoy constitutional protections such as free speech or free assembly.

Today, 70 percent of all international trade is directed by 500 corporations. The largest 300 corporations control about one-fourth of all the goods-producing assets in the world. The largest 100 have incomes greater than half the countries of the world.

We the people have to figure out how to exercise our sovereign authority to stop corporations from poisoning and destroying our communities, and to get corporations to serve the common good. To accomplish this, we can do the following:

- Take away the constitutional privileges and immunities which corporations use to rule the land.

> - Recharter cooperating corporations on *our* terms, with people in our communities writing and enforcing the new rules.
> - Remove the most abusive, most anti-democratic corporations by revoking their charters.
>
> —*Richard Grossman*

followed up with similar reports and editorials. In his article, Schneider wrote that dioxin, "once thought to be much more hazardous than chain smoking, is now considered to be no more risky than spending a week sunbathing," attributed this statement to "some experts." Schneider later admitted that he made up the comparison himself (Monks, 1993).

Schneider's campaign against dioxin was critiqued and documented in detail in a 1993 *American Journalism Review* article by Vicki Monks. According to Monks, Schneider said that his coverage of dioxin was "forcing environmental journalists to look at how they are writing these stories...and whose side they are taking." Schneider went on to say that he's at the "forefront" of a "new area of environmental reporting" in which "we look [at] and view all sides equally skeptically, and that we come to conclusions based on data, not the frantic ravings of one side or another."

"Unfortunately," wrote Monks, "many news organizations made conclusions based on Schneider's assertions, such as the 'fact' that dioxin is no more dangerous than sunbathing, without reviewing the data or checking the sources of the information."

Vernon Houk was the source of many news articles disclaiming the dangers of dioxin. In his role at the CDC, Houk evaluated health risks in many contaminated communities around the country and was involved in the decision in 1983 to evacuate the residents of Times Beach. Houk was also responsible for derailing the CDC Agent Orange study.

Houk was quoted in a 1991 front-page *New York Times* article as saying that he had made a mistake in urging the federal government to evacuate Times Beach, and that "many scientists now believe dioxin isn't as bad as we thought." Houk originally made this statement at a conference sponsored by Syntex, the company responsible for the cleanup of Times Beach. In August of that same year, Houk told the *Seattle Times* that he believed the pulp and paper industry in the Northwest "has reduced dioxin levels enough to protect public health." Houk, who died in 1994 at age sixty-five of lung cancer, fashioned a new career for himself as the dioxin-is-not-so-bad spokesman for industry. He even admitted that in his proposal for relaxed dioxin standards, he copied word-for-word reports given to him by the paper industry (*Waste Not,* 1991).

According to Monks, "several newspapers with direct financial interests in paper and timber companies have taken editorial positions supporting relaxed dioxin standards without disclosing their ties to the industry." As examples, Monks cites Central Newspapers, which includes the *Arizona Republic* and *Indianapolis Star,* owned by former vice president Dan Quayle's family; the Times Mirror Company, which owns the *Los Angeles Times*; the *Chicago Tribune*; the *Washington Post*; and the *New York Times* (Monks, 1993).

Another version of the "not sure yet" argument is "since we aren't sure of all the sources of dioxin, it would be unfair to pick on the ones we do know about." That's like saying that until we know all the sources of airplane crashes, we don't have any responsibility to prevent those crashes we know how to prevent. Yet another version claims that "too much of what we know is from rats and guinea pigs. We haven't seen the same effects on people, so people must respond differently." Are they suggesting that we must do controlled experiments on people before we can be sure?

Your educational efforts to counter these arguments should stress what we do know and why our present knowledge is enough to warrant action. Emphasize that there's a difference between the scientific need for proof and the precautionary principle. Certainty or proof in science would require that the same study be done over and over again, each time arriving at the same precise result.

Absolute scientific certainty can't be the standard for public health protection. Dioxin-polluting corporations will never be certain enough. The tobacco industry still doesn't believe that smoking causes cancer.

Environmental decisions should be made so that the burden of proof is placed on showing that a chemical or practice is safe, not is harmful. Rather than using all chemicals until they are proven harmful, we should demand that all chemicals be shown to be safe before they are used. The United States-Canadian International Joint Commission on the Great Lakes has adopted this precautionary principle of showing safety first, recommending that "the concept of reverse onus, or requiring that a substance is not toxic or persistent before use, should be the guiding philosophy of environmental management agencies in both countries."

"We Know Plenty and There's Nothing to Worry About"

Sometimes corporations and government will explain that dioxin really isn't that bad for you. Agent Orange in Vietnam and the dioxin accident in Seveso, Italy, have only caused chloracne and temporary nervous system damage, they will argue.

Arguing along similar lines, the French Academy of Sciences issued a stinging rebuttal to the EPA report. The Academy points out that no fatal case of poisoning by dioxin-contaminated products has ever been reported. "Man appears to be much less sensitive to consequences of exposure to PCDD/F (dioxin) than most animal species studied," the academy asserted.

The Chemical Industry Institute of Technology (CIIT) argues that just because dioxin at or near the current environmental levels of exposure may bind with Ah receptors and alter gene expression, we shouldn't assume that it harms human health. The industry is sure to argue that the amounts of dioxin released into the environment are so small that they must be harmless. Table 17-1 shows that very small amounts can have huge impacts.

Be prepared for arguments that natural dioxin is formed in forest fires and wood stoves. According to the CCC's Kip Howlett, "Mother Nature carries the starter kit for dioxin. Do we bring Smokey the Bear to the witness box?"

Table 17-1

Impact of One Part Per Billion

Substance	One ppb in water* would be enough to:
Chromium	Plate 50,000 car bumpers
Gold	Run the federal government for nearly 20 minutes or support 50 average families for eternity
Herbicides	Kill all of the dandelions in 100,000 lawns
Insecticides	Fill 5,000,000 aerosol cans of bug killer
Lead	Cast 1,000,000 bullets
Mercury	Fill 4,000,000 rectal thermometers
Phenols	Produce 250,000 bottles of Lysol

The amount of water used in the Los Angeles area in one year.
Source: Chriswell (1977)

The argument that dioxin is "natural," and is formed without the presence of man-made chemicals, is made in efforts to minimize and dismiss concerns about the toxicity of dioxin. "If dioxin has been around since the days of the cave men, then it can't be that bad." But, as discussed in the Introduction, time-dated dioxin levels identified in lake sediments, Chilean mummies, and other sources argue against this theory. While some minute levels of dioxin may have been evident in prehistoric times, the fact is that levels today are much higher and that substantial levels did not start showing up until after the modern chemical industry began in the 1940s.

Despite this evidence, arguments that forest fires account for some amount of dioxin formation continue to be made. For example, an editorial in the magazine *Science* by former editor Philip Abelson stated that "Anthropogenic (man-made) production of dioxin has decreased during the past two decades and is smaller than that created by combustion of wood." Table 3-2 shows that wood burning accounts for only 4 percent of dioxin emissions, so Dr. Abelson is clearly mistaken. Later in the same editorial, Abelson says, "Two research groups have concluded that forest and brush fires are major sources of PCDDs (polychlorinated dioxins)." If one examines the two references that he quotes, they say nothing of the sort. The first reports that dioxin could come from burning wood but did not estimate the total emissions from forest fires. The second reference estimates that as much as 60 kilograms of dioxin comes from forest fires in Canada and 25 to 110 kilograms from other sources (Abelson, 1994). Clearly, this reference does *not* say that the amount from other sources is smaller. And in any case, 60 kilograms of dioxin is more than six times the total dioxin released from all U.S. sources combined—an exaggeration beyond belief. Corporate spin doctors will argue that the EPA has overstated the risk posed by dioxin. Even if the EPA tripled their safe dose estimates, the average intake of dioxin would be 94 times the EPA's recommendation over a lifetime.

Corporate spokespeople will also argue that since the average dioxin level in people's fatty tissues is lower than it used to be, the problem has already been solved. Your response to these arguments is to counterpunch with the truth. Cite all the studies in this book that show the health effects of dioxin. Tell the stories of how Monsanto and Dow created fraudulent studies to hide the truth about the dangers of dioxin. Question the veracity of other reports paid for by industry.

Dioxin levels are probably going down because PCBs were banned in 1978, and the herbicide 2,4,5-T was removed from market in 1977. But levels are still far too high. Wood fires produce dioxin because dioxin fallout has contaminated the wood, just as house fires produce dioxin because of burning PVC.

"Chlorine's Benefits Outweigh Any Dioxin Risk"

The corporations will argue that their products save lives and protect society. Whatever harm dioxin may cause, they say, is less than the damage that would be done to society and industry by stopping its production. Risk assessment numbers will be thrown out to show the risk is only one in 10,000 or one in 100,000. These numbers will be weighed against the widespread benefits of chlorine-based products.

Risk assessment is the scientific-sounding name for corporations and their friends in government deciding what is an acceptable amount of poison for you. Risk assessment is based on the false premises that we are exposed to only one toxic at a time, that we have never been exposed before, and that cancer is the only health problem we need to consider.

Oregon historian and activist Carol Van Strum suggests that we counter risk assessment arguments with the principle of informed consent. You do not need a Ph.D. to say no to dioxin exposure. You simply need to say, "No," and challenge an industry's or a CEO's right to expose you without your consent. Your weapon here is the age-old common law on assault and battery: Mr. Incinerator CEO could be prosecuted for assault and battery if he *kissed* you without your consent and against your will, and even if he sincerely believes that dioxin is good for you, he has no more right to put it inside your body without your consent than he has to kiss you.

People understand this principle, the principle that "your rights end at the tip of my nose, buddy." They do not need science or government to assert this principle. They simply need to say no, and point out that no amount of permits or registrations or licenses gives Mr. CEO the right to put this molecule in their bodies without their consent. Organizing around the issue of informed consent/assault and battery cuts through all the risk assessment and scientific quibbling—indeed, it uses risk assessments against industry, because risk assessment is an *admission* that people will be exposed and that some of them will die.

"Personal Responsibility, Yes; Corporate Accountability, No"

This argument is based on the rationale that since we get almost all of our dioxin from food, dioxin exposure is a personal diet issue, not a corporate or government issue. A corporate spokesperson may explain, "The air inside this incineration facility is safer than ice cream, so we don't have to clean it up. To stop dioxin exposure, the American people should just eat a well balanced, low-fat diet." Responsibility for dioxin exposure is shifted from products to consumers, and from corporations to individuals.

To make the corporations accountable, activists must argue that we have the right to safe food. Focus on the corporations that created the dioxin fallout that got onto the grass, and then into the cow, and then into the ice cream. (Borden actually makes both dairy products and PVC products like food wrapping) Dioxin contamination is not the cow's fault. It's not the farmer's fault. And it's not the fault of the person who's eating the ice cream. The buck should stop where the big bucks are being made on dioxin—at the source.

"We'd Like to Stop Making Dioxin, But It Is Just Too Expensive"

The Chlorine Chemistry Council, quoting a study by Charles River Associates (CRA), estimates that the cost of finding chlorine substitutes and converting U.S. plants to use them would exceed $60 billion. This means that each one of the 367,000 jobs that CRA says depend directly on chlorine would cost $163,488 to convert. That estimate sounds a whole lot like the utilities' estimate of the price tag for reducing emissions of sulfur dioxide, the chief ingredient of acid rain. Utilities warned that the cost of cleaning up their emissions would be $1,000 a ton. The real cost turned out to be $140 a ton (Carey, 1995).

Your Stop Dioxin Exposure Campaign should focus on demonstrating that the cost of change is a much better investment than the cost of the status quo. The cost of the status quo must include the cost to our health. What is the price of infertility? Of birth defects? Of cancer?

Alternatives *do* exist for all the major sources of dioxin poisoning. Your coalition's job is to educate communities about the alternatives, and then to organize to make the alternatives irresistibly attractive. Make sure that the alternatives include jobs and job protections. When you are actually asking corporations to make big investments in new technologies, consider the use of industrial revenue bonds and other public incentives to help pay for the capital investments and to protect workers' jobs.

"Poisoning is the Cost of Civilization"

According to this argument, if we don't want dioxin, we must be ready to give up cholera-free water, television, affordable housing, automobiles, and life-saving pharmaceuticals. Your coalition can counter with the success of alternative technologies.

Dow Chemical, the world's biggest maker of chlorine, is already developing new products that will position the company to profit from a chlorine phase-out. Among Dow's new products are:

- a chelating agent that pulls out the heavy metals that build up in chlorine-free pulp bleaching,

- a water-based cleaning system that can be used as an alternative to chlorinated solvents,

- biological "crop-protection" products to replace some chlorinated pesticides, and

- a special line of chlorine-free plastics, which company officials say will replace PVC.

After you respond to the doomsday scenario of life without chlorine, tell the real stories of people who are being hurt by dioxin. How do you tell a child sick with dioxin-related sarcoma, or a couple unable to have children, that their tragedy is the cost of civilization? Tell the stories that reveal the hidden costs of human suffering.

Contributors

Charlotte Brody
Stephen Lester, M.S.
Charlie Cray
Richard Grossman

Chapter Eighteen

Bruce, Martha and the Vandellas, Felix, Dioxin, and You

The first half of this book describes what the dioxin polluters are doing to us. Infertility. Endometriosis. Impaired immune systems. Diabetes. Decreased testes size. Liver damage. Birth defects. Cancer. The second half describes what we can do to them to make them stop.

And we must make them stop. No new mother should have to suffer the agony of knowing that her breast milk is filling her child with the dioxins that have polluted her over her lifetime. And no corporation has the right to contaminate a mother's milk, or any of our food, water, or air.

It takes a leap of faith to get things going
It takes a leap of faith, you got to show some guts.

—Bruce Springsteen

It takes a leap of faith to move from understanding the first half of this book to acting on the second half. And leaps of faith, as Bruce Springsteen writes, take guts. It isn't easy to make that tenth phone call, knock on a stranger's door, or take on a multinational corporation. But it is absolutely necessary.

We can stop dioxin exposure. Slavery was abolished. Child labor ended. Women won the right to vote. Labor won the right to organize. Segregation of public facilities was ended. The War in Vietnam was stopped. DDT was banned. Countries stopped above-ground nuclear testing. Families won the right to manage their own fertility. The Berlin Wall came down. Apartheid ended. Each of these societal leaps resulted from the leaps of faith of ordinary people who joined together in organized and persistent efforts to demand and win change.

It is now time to apply the lessons of other social justice movements to stop the dioxin poisoning of the American people. We must build a movement that not only wins a series of local and national reforms, but stays around to guard and improve on those victories. Dioxin isn't the only environmental source of our health problems. We will have other battles to fight. Our movement won't be strong enough to stop dioxin exposure and won't be permanent enough to protect what we've won unless we restore our democracy. Our political life cannot be limited to voting and writing checks for our favorite direct-mail causes. In every community, we must restore democratic public life.

For the people to regain control of democratic institutions, corporations will have to relinquish some of their power. For too many years, corporations have dictated to and threatened the American people, and have bought their way into Congress, state capitols, and local governments. Corporations have benefitted handsomely while their workers and the public have made huge health, environmental, and economic sacrifices. As our efforts to stop dioxin and rebuild democracy begin to restore the balance between corporate power and public health, the money corporations now spend on lobbying and electoral campaigns can be spent on creating and maintaining industrial practices that don't make people sick.

To change the behavior of corporations and institutions, we need to affect their bottom lines. We need to make it more expensive to pollute than it is to change. Corporations will not change because CEOs wake up one morning and decide that stopping dioxin pollution is the right thing to do. Corporations will change because people—consumers, voters, and workers—convince them that they must change.

This change will not come from Washington, D.C. It will not appear in the form of a top-down regulatory mandate. Changes in corporate behavior will only be accomplished through people working at the local level, then joining together at the state level, and then at the national level. Change depends on you, me, and millions of others who are willing to make that leap of faith from education into collective action.

Nowhere to run to, nowhere to hide.

—Martha and the Vandellas

Eliminating the sources of dioxin or avoiding additional exposures are not within an individual's control. Martha and the Vandellas might as well have been singing about dioxin. No mountain is high enough, no yogurt is organic enough to protect us from dioxin.

There are small steps that individuals can take to reduce dioxin exposure, such as eating less beef and dairy foods and not using chlorinated pesticides. Yet there is nothing individuals alone can do, through a change of lifestyle or diet, to completely protect themselves or their children from dioxin exposure. There's nowhere to run but to each other.

In a democracy, the highest office is the office of citizen.

—Supreme Court Justice Felix Frankfurter

Take your first step toward stopping dioxin exposures today by talking with one person at school, work, church, your child's soccer game, or your exercise class. Find a way to talk about how dioxin is affecting both of you and what you can do about it. As you talk about stopping dioxin exposure, you'll be taking a first step in rebuilding our democracy.

Democracy is defined in Webster's dictionary as "a government in which the power is vested in all the people." The Stop Dioxin Exposure Campaign can be a vehicle for us to take back the power that belongs to us, and begin to rebuild our fragile democracy.

We need to take on this challenge to protect future generations. We need to take on this challenge to reclaim the power that should rightly be vested in all the people. The job is too big for some national organization or remote coalition to achieve on our behalf. To reclaim our health and rebuild our democracy we must all play a role. Our country's power is vested in the people, and the people must act.

Afterword

This book is only a first chapter. CCHW wants to provide opportunities for collaboration and connection between grassroots efforts to stop dioxin exposure around the country. We want to let people know what strategies are working in communities and what people have figured out about building new democratic structures. One way we're doing this is through the information superhighway. CCHW has established a dioxin bulletin board so you can let other activists know what you're doing and find out what's going on around the country. CCHW will also be producing on-line updates to this book. You can become part of CCHW's Stop Dioxin Exposure Campaign bulletin board by sending the e-mail message: subscribe dioxin-l <your name> to listproc@essential.org or contact CCHW at CCHW@essential.org.

We're also available to talk to people on the telephone, write and send letters through the mail, and visit different communities. We'll be sending out campaign updates to everyone who lets us know that they want to be a part of this effort. You can also call or write us at the address and telephone number below to order t-shirts and buttons: Stop Dioxin Exposure Campaign, CCHW, P.O. Box 6806, Falls Church, VA 22040. (703) 237-2249.

Appendix A

The Chemistry of Dioxin and Dioxin-like Substances

Even if you hated high school chemistry, try reading this appendix. You might find that intimidating names such as 2,3,7,8-tetrachlorodibenzodioxin begin to make a little more sense.

The basic building block for dioxin-like chemicals is benzene, which contains 6 carbon atoms and 6 hydrogen atoms linked up in a ring. (See Figure A-1). For shorthand, the carbons and hydrogens are usually omitted, and the ring is depicted as shown in Figure A-2.

Dioxin and dioxin-like chemicals are composed of two benzene rings hooked together in one of 3 different ways. (See Figure A-3). If they are hooked together by a 6-member ring containing 2 oxygens, they belong to the family of di benzo di oxins (di for 2, benzo for benzene rings, di for two, and oxin for oxygen). If they are hooked together by a 5-member ring containing 1 oxygen, they are called furans (di benzo furan). If they are hooked together directly, they are called bi phenyls, which are the basic building blocks of PCBs. The dioxins and furans have 3 rings in their structure, but the bi phenyls have only 2. (See Figure A-3.)

Let's move on to the chlorines. Each of the hydrogen atoms in the ring can be removed, and a chlorine atom substituted. To

Figure A-1

Benzene Ring With All the Atoms

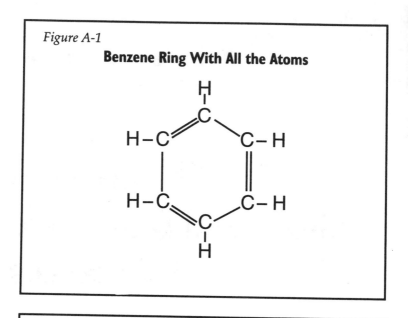

Figure A-2

"Shorthand" Benzene Ring Without Showing Atoms

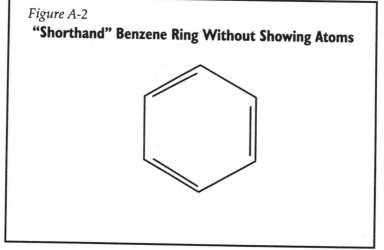

keep track of where the chlorines are, the molecules are numbered as shown in the figures above, and the name of the molecule gives the location of the chlorines. For example, 2,3,7,8-tetra chloro di benzo di oxin has 4 chlorines at positions 2,3,7, and 8. (See Figure A-4.)

Figure A-3
Diagrams of Dioxin, Furan, and Biphenyl

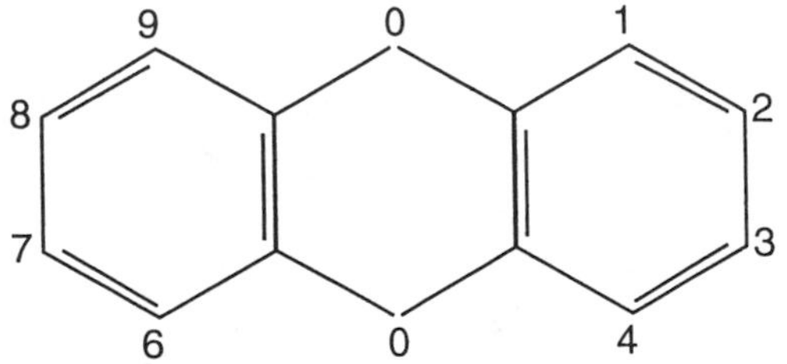

Dioxin

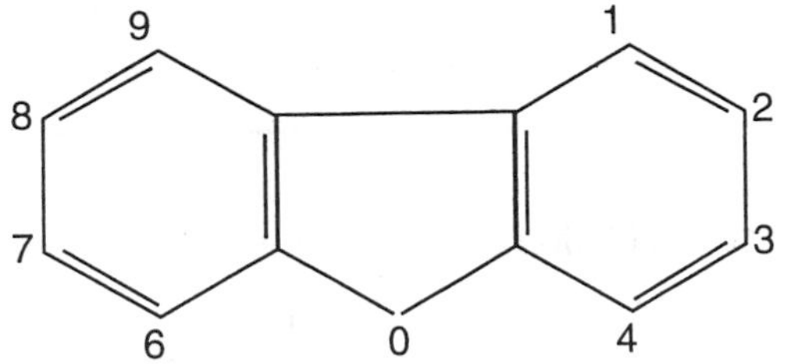

Furan

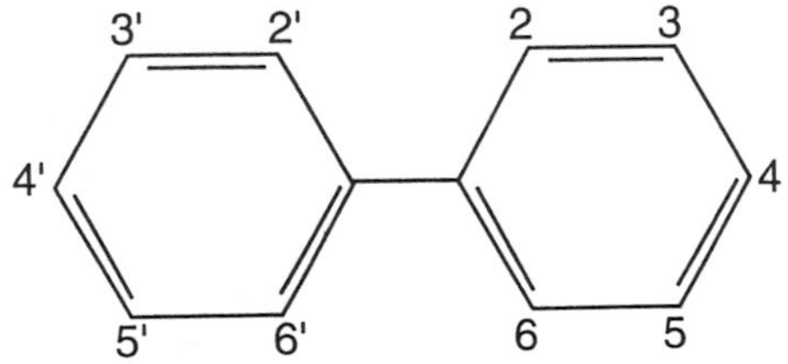

Biphenyl

Figure A-4

Diagram of 2,3,7,8-TCDD

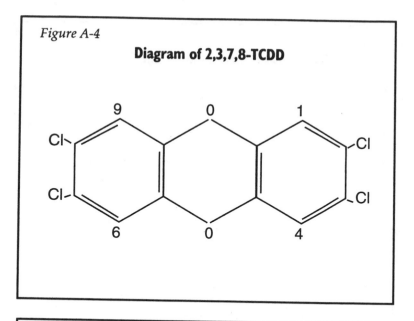

Table A-1

Number of Different Arrangements of Chlorines

	Dioxins	Furans	Biphenyls
1. mono	2	4	3
2. di	10	16	12
3. tri	14	28	24
4. tetra	22	38	42
5. penta	14	28	46
6. hexa	10	16	42
7. hepta	2	4	24
8. octa	1	1	12
9. nona	0	0	3
10. deca	0	0	1
TOTAL	75	135	209

The number of chlorines (chloro) is indicated by the prefixes listed in Table A-1. This table also shows the number of possible molecules that can be formed, depending on the number and arrangement of chlorines. For example, 2 chlorines (di chloro) can arrange on the dioxin molecule in 10 different ways, so there are 10 different di chloro di benzo di oxins. All together, there are 75 different dioxins, 135 different furans, and 209 different PCBs.

Some definitions will help us talk about the array of related chemicals.

- *Congener:* A specific member of the same large family of compounds. There are 75 congeners of chlorinated dibenzodioxins; 135 of dibenzofurans; 209 of PCBs.

- *Homologue:* One of a group of similar compounds with the same number of chlorines. There are 10 homologues of dichlorodibenzodioxins, 16 of dichlorodibenzofurans, and 12 of PCBs.

- *Isomer:* A specific compound that belongs to a homologous group. 2,3 dichloro dibenzodioxin is an isomer of the dichlorodibenzodioxins.

Bromine is a chemical closely related to chlorine, and the same set of chemicals can exist with bromine in the place of chlorine. For example, in Michigan, the fire retardant PBB (poly brominated biphenyl) accidentally got into animal feed and contaminated cows and milk. Both the chlorinated and brominated chemicals are toxic, but the chlorinated ones are more common, so most people hear more about those.

Not all of these chemicals are equally toxic. Only 7 of the 75 dioxins, 10 of the 135 furans, and 11 of the 209 PCBs have dioxin-like toxicity. One key factor in toxicity is the shape of the chemical, and this is determined by how many chlorines it contains, and where the chlorines are attached.

The relationship of shape to toxicity is probably due to the fact that shape determines binding to the receptor molecule, and binding to the receptor appears to be necessary for toxicity. For example, certain PCBs that tend to be flat in space with both of the rings in the same plane are called coplanar, and these are much more toxic than the others.

Contributors

Dr. Beverly Paigen
Stephen Lester, M.S.

Appendix B

Conversion Charts

Table B-1
Converting from One Parts Per Unit to Another

Original Unit	New Unit			
	ppm	ppb	ppt	ppq
ppm		1,000	1,000,000	1,000,000,000
ppb	.001		1,000	1,000,000
ppt	.000001	.001		1,000
ppq	.000000001	.000001	.001	

ppm = parts per million
ppb = parts per billion
ppt = parts per trillion
ppq = parts per quadrillion

Table B-2

Conversion Factors for Concentrations in Solids

This table can be used to convert many of the different units used in this book. To use this chart, look down the column on the left to find the denominator of the unit you want to convert (the denominator of mg/kg is kg; the denominator of ng/mg is mg). Then go across the row to the original unit (mg/kg). Next go to the bottom half of the column to find the appropriate conversion factor for whichever unit you wish to convert (1 ppm).

Denominator	Original Units				
kilogram (kg)	g/kg	mg/kg	ug/kg	ng/kg	pg/kg
milligram (mg)	ug/mg	ng/mg	pg/mg	fg/mg	ag/mg
microgram (ug)	ng/ug	pg/ug	fg/ug	ag/ug	
nanogram (ng)	pg/ng	fg/ng	ag/ng		
picogram (pg)	fg/pg	ag/pg			
femtogram (fg)	ag/fg				
attogram (ag)					
	Conversion Factors				
	1000 ppm	1 ppm	.001 ppm	.000001 ppm	.000000001 ppm
	1000000 ppb	1000 ppb	1 ppb	.001 ppb	.000001 ppb
	1000000000 ppt	1000000 ppt	1000 ppt	1 ppt	.001 ppt
	1000000000000 ppq	1000000000 ppq	1000000 ppq	1000 ppq	1 ppq

A p p e n d i x   C

Sample "Bad Boy" Ordinance

The following is a by-law prohibiting and/or regulating contracts between the town of Palmer, Massachustetts, and business entities, persons, or employees of business entities who have been convicted of bribery, attempted bribery, collusion, restraints of trade, and price-fixing.

Section I

No person or business entity shall be awarded a contract or subcontract by the Town of Palmer if that person or business entity:

(a) Has been convicted of bribery or attempting to bribe a public officer or employee of the Town of Palmer, the State of Massachusetts, or any other public entity, including, but not limited to the Government of the United States, any state, any local government authority in the United States in that officer's or employee's capacity; or

(b) Has been convicted of an agreement or collusion among bidders or prospective bidders in restraint of freedom of competition by agreement to bid a fixed price, or otherwise; or

(c) Has made an admission of guilt of such conduct described in paragraphs (a) or (b) above, which is a matter of record, but has not been prosecuted for such conduct, has made an admission of guilt of such conduct which term shall be construed to include a plea of nolo contenders.

Section 2

A person, business entity, officer or employee of a business entity, convicted of one or more of the crimes set forth in Section 1, shall be ineligible for the awarding of a contract or subcontract by the Town of Palmer for a period of three years, following such conviction or admission in the case of an admission of guilt of such conduct, which is a matter of record, but which has not been prosecuted.

Section 3

For purposes of this By-law, where an official, agent or employee of a business entity has committed any offense under this By-law; as set forth in Sections 1 or 2, on behalf of such an entity and pursuant to the direction or authorization of an official thereof (including the person committing the offense, if he is an official of the business entity), the business entity shall be chargeable with the conduct hereinabove set forth. A business entity shall be chargeable with the conduct of an affiliated entity, whether wholly owned, partially owned, or one which has common ownership or a common Board of Directors. For purposes of this Section, business entities are affiliated if, directly or indirectly, one business entity controls or has the power to control another business entity, or if an individual or group of individuals controls or has the power to control both entities. Indicia of control shall include, without limitation interlocking management or ownership, identity of interests among family members, shared organization of a business entity following the ineligibility of a business entity under this Section, using substantially the same management, ownership or principals as the ineligible entity.

Section 4

Any party who claims that this By-law is inapplicable to him, them, or it, because a conviction or judgement has been reversed by a court with proper jurisdiction, shall prove the same with documentation satisfactory to Town Counsel.

Section 5

The Town of Palmer shall not execute a contract with any person or business entity until such person or business entity has executed and, filed with the Town Clerk an affidavit executed under the pains and penalties of perjury that such person or entity has not been convicted of any violation described in Section 1 paragraphs (a) or (b), and has not made an admission of guilt or nolo contenders as described in Section 1, paragraph (c). In the case of a business entity such affidavit shall be executed by, in the case of a partnership, the general partner(s), and in the case of a corporation, the president.

Section 6

All By-laws or parts of By-laws in conflict herewith are hereby repealed.

Section 7

Should any provision or section of this By-law be held unconstitutional or invalid, such holding shall not be construed as affecting the Validity of any of the remaining provisions or sections, it being the intent that this By-law shall stand notwithstanding the invalidity of any provision or section thereof.

Principles Of Environmental Justice

The following declaration of principles was adopted at the First National People of Color Environmental Leadership Summit, which took place October 24–27, 1991, in Washington, D.C.

Preamble

We, the people of color, gathered together at this multinational People of Color Environmental Leadership Summit, to begin to build a national and international movement of all peoples of color to fight the destruction and taking of our lands and communities, do hereby re-establish our spiritual interdependence to the sacredness of our Mother Earth; to respect and celebrate each of our cultures, languages and beliefs about the natural world and our roles in healing ourselves; to insure environmental justice; to promote economic alternatives which would contribute to the development of environmentally safe livelihoods; and, to secure our political, economic and cultural liberation that has been denied for over 500 years of colonization and oppression, resulting in the poisoning of our communities and land and the genocide of our peoples, do affirm and adopt these Principles of Environmental Justice:

- Environmental justice affirms the sacredness of Mother Earth, ecological unity and the interdependence of all species, and the right to be free from ecological destruction.

- Environmental justice demands that public policy be based on mutual respect and justice for all peoples, free from any form of discrimination or bias.

- Environmental justice mandates the right to ethical, balanced and responsible uses of land and renewable resources in the interest of a sustainable planet for humans and other living things.

- Environmental justice calls for universal protection from nuclear testing and the extraction, production and disposal of toxic/hazardous wastes and poisons that threaten the fundamental rights to clean air, land, water and food.

- Environmental justice affirms the fundamental right to political, economic, cultural and environmental self-determination of all peoples.

- Environmental justice demands the cessation of the production of all toxins, hazardous wastes, and radioactive materials, and that all past and current producers be held strictly accountable to the people for detoxification and the containment at the point of production.

- Environmental justice demands the right to participate as equal partners at every level of decision-making including needs assessment, planning, implementation, enforcement and evaluation.

- Environmental justice affirms the right of all workers to a safe and healthy work environment, without being forced to choose between an unsafe livelihood and unemployment. It also affirms the right of those who work at home to be free from environmental hazards.

- Environmental justice protects the right of victims of environmental injustice to receive full compensation and reparations for damages as well as quality health care.

- Environmental justice considers governmental acts of environmental injustice a violation of international law, the Universal Declaration On Human Rights, and the United Nations Convention on Genocide.

- Environmental justice must recognize a special legal and natural relationship of Native Peoples to the U.S. government through treaties, agreements, compacts, and covenants affirming sovereignty and self-determination.

- Environmental justice affirms the need for an urban and rural ecological policies to clean up and rebuild our cities and rural areas in balance with nature, honoring the cultural integrity of all our communities, and providing fair access for all to the full range of resources.

- Environmental justice calls for the strict enforcement of principles of informed consent, and a halt to the testing of experimental reproductive and medical procedures and vaccinations on people of color.

- Environmental justice opposes the destructive operations of multi-national corporations.

- Environmental justice opposes military occupation, repression and exploitation of lands, peoples, and cultures, and other life forms.

- Environmental justice calls for the education of present and future generations which emphasizes social and environmental issues, based on our experience and an appreciation of our diverse cultural perspectives.

- Environmental justice requires that we, as individuals, make personal and consumer choices to consume as little of Mother Earth's resources and to produce as little waste as possible; and make the conscious decision to challenge and reprioritize our lifestyles to insure the health of the natural world for present and future generations.

Sample Good Neighbor Agreement

This agreement is between Lewcott Corporation, a Massachusetts corporation (hereafter called "Lewcott"), and the Quinsigamond Village Health Awareness Group, an unincorporated association (hereinafter called QVHAG).

Lewcott and QVHAG agrees as follows:

I. Odors

A. Odors from current phenolic based coatings and future water based coatings (hereinafter called the "odors") shall not be emitted from the Lewcott facility at 280 Greenwood Street, Worcester, MA.

B. In the event that odors are emitted and written notice is given to Lewcott of the detection of odors by one or more neighbors and DEQE [Department of Environmental Quality Engineering] verifies that:

1. In reference to phenolic odors, the Consent Order was violated; or

2. With reference to odors from future water based coating an odor exists through a review of test data or on site direction;

Lewcott shall immediately cease production on the problematic line and not resume production until:

1. the cause has been identified;

2. corrective actions have been identified and implemented; and

3. Lewcott receives written approval from the DEQE and Lewcott provides copy of said approval to QVHAG.

C. A violation of this odor prohibition would be considered to have occurred when an odor is reported by a resident and a representative of DEQE verifies (i) with reference to phenolic odors that the Consent Order was violated or (ii) with reference to odors from future water based coatings that an odor exists through a review of test data or on-site detection. As used in this agreement "Consent Order" means that certain Modified Consent Order dated March 6, 1987, between the Department of Environmental Quality Engineering and Eli Sandman Company, a subsidiary of Lewcott.

2. Flammable Storage License

A. Lewcott agrees to surrender the flammable storage license issued to it by the Worcester License Board on November 15, 1987.

B. Lewcott agrees that it shall not make any request to the Worcester License Board for any extensions of such license beyond November 15, 1987, or apply at any future time to the Worcester License Board for such a license to store flammable substances at 280 Greenwood Street.

3. Deadlines

A. Phenolic Line. Lewcott agrees to cease phenolic coating at 280 Greenwood Street not later than October 31, 1987.

B. All Coating Operations. Lewcott agrees to cease all coating operations at 280 Greenwood Street not later than July 31, 1988.

C. Cutting Operations. Lewcott agrees to cease its cutting operations at 280 Greenwood Street not later than July 31, 1988.

D. Final Relocation. Lewcott agrees to have completed relocation of all remaining company operations and offices from 280 Greenwood Street not later than October 31, 1988.

E. Good Faith Expeditiousness. Lewcott agrees to exercise good faith in ceasing and relocating operations earlier than the dates specified in this Agreement as proves feasible.

F. No entity affiliated with Lewcott or BBF Corp. will occupy or own said premises after October 31, 1988.

G. If the Lessor of the premises does not agree to-the amendment referred to in paragraph 4.B below, Lewcott agrees that it will not sublease said premises to any entity engaged in a coating, plating or chemical business, unless QVHAG approves.

4. Enforcement

A. Any violation of this Agreement by Lewcott shall result in a penalty against Lewcott of $5,000 per day of violation payable within 10 days of said violation to a community charity designated by QVHAG. In the event that Lewcott is assessed administrative or other penalties for the same violation under the Consent Order, the same shall be credited against the penalty due hereunder. In addition to the award of these mandated penalties, the arbitrator may issue other appropriate orders to either party to comply with the terms of this Agreement.

B. If a dispute arises with any aspect of the Agreement which cannot be amicably resolved, Lewcott and QVHAG hereby agree to submit such dispute to arbitration on the following terms. Upon written demand of either party the dispute shall be referred to an impartial arbitrator mutually agreeable to both parties. If within 20 days of the demand, the parties are unable to agree to an impartial arbitrator, either party may submit the dispute to the

American Arbitration Association for arbitration. All arbitration hereunder shall be in accordance with the commercial arbitration rules of said Association. The costs and expenses of the arbitration, including the reasonable attorneys' fees of the parties, shall be paid as determined by the arbitrator except as provided in the following sentence. If the arbitrator determines that the issues raised by (i) QVHAG are frivolous, not-in good faith and without basis all of said costs and expenses shall be paid by QVHAG or (ii) Lewcott are frivolous, not in good faith and without basis all of said costs and expenses shall be paid by Lewcott. Judgement upon the award and any order rendered by the arbitrator may be entered in any court having jurisdiction thereof. A decision of the arbitrator shall be final and binding. Said decision must be obeyed within ten days. In the event that such decision is not complied with within ten days and court action becomes necessary, the level of penalties awarded in a court enforcement action shall be three times the penalty stipulated in this Agreement. In such a court action, reasonable attorney and expert fees shall be awarded to the prevailing party.

C. Arbitration under paragraph 4.B hereof shall be the sole and exclusive remedy of any dispute referred to in said paragraph.

D. Lewcott agrees to present to the lessor of the 280 Greenwood Street premises an amendment to the lease incorporating the provision of paragraphs 1, 2, 3, and 5.A of this Agreement, and terminating the Lease effective 10/31/88, such amendment to be in form annexed hereto as Exhibit A with appropriate insertions for names of parties, dates of lease, paragraph numbers, etc.

E. Lewcott agrees to allow QVHAG, at reasonable times and intervals, the right to on-site visual verification of Lewcott's compliance with this Agreement.

5. Other Terms and Conditions

A. Unforeseeable Circumstances. Lewcott shall be bound by the terms of this Agreement except for unforeseeable circumstances such as Acts of God which prevent or delay its implementation. In the event of an instance creating such circumstances, Lewcott would notify QVHAG immediately upon the occurance of such an incident and define to the best of its ability the extent of delay or noncompliance that will result. Delays in the implementation of clauses A, B and C of paragraph 3 shall be limited to an aggregate of 45 days unless the parties agree otherwise. If Lewcott requests and QVHAG does not grant an extension beyond the 45 day aggregate, the question of whether such an extension should be granted shall constitute a dispute subject to resolution by either party under the arbitration provisions of paragraph 4.B of this Agreement and the arbitrator may grant all or part or none of the requested extension up to an additional 30 days. The compliance date in clause D of paragraph 3 is not subject to extension for any reason.

B. Removal of Signs Along Greenwood Street. QVHAG agrees, to the best of its ability, to remove signs related to Lewcott which its members have posted along Greenwood Street. In the event of any significant violation of the Agreement, QVHAG reserves the right to begin reposting of the signs.

C. All notices hereunder shall be sufficient if in writing and mailed, postage pre-paid, certified mail return, receipt requested, and addressed as follows: if to Lewcott, to it at 280 Greenwood Street, Worcester, MA 01607; and if to QVHAG, to it as P. 0. Box 1715, Sta. C, Worcester, MA 01607. Either party may change its address by like notice given at least 10 days prior to the change effective date.

D. Any notice, agreement or waiver made hereunder by Cynthia Phillips shall be binding upon QVHAG for all purposes.

6. Ratification

This Agreement will become effective if ratified by a majority vote at a community meeting conducted by QVHAG. Executed this 20th day of March, 1987 by duly authorized officers.

—from The Citizens Toxics Protection Manual,
National Campaign Against Toxic Hazards

Appendix F

Sample Dioxin Resolution

This resolution was passed by the Midway, Texas, High School PTA and sent to the full delegate body of the Texas PTA Board of Directors for consideration at the November 1995 convention.

Midway High School Dioxin Resolution

Whereas, Dioxin, the most toxic substance created by humans, is formed as an accidental by-product in numerous industrial processes involving chlorine.

Whereas, Dioxin is persistent in the environment, food chain, and in our bodies.

Whereas, Dioxin is a by-product of waste incineration containing chlorine, chemical and plastics manufacturing, paper and pulp bleaching, burning hazardous wastes in cement kilns.

Whereas, Dioxin is cross generational, passing from mother to child through the placenta, and via mother's milk.

Whereas, the creation of Dioxin is an avoidable hazard creating numerous adverse health effects such as cancer, hormonal disruption, infertility, impaired child development, suppression of the immune system, endometriosis, and diabetes.

Whereas, children bear the highest exposures, current levels of dioxin in the bodies of the general population are already in the range at which health effects are known to occur in laboratory animals, according to EPA.

Whereas, some communities are subject to even greater exposures and health risks because of disproportionate siting of polluting facilities in minority communities.

Whereas, wildlife are now suffering severe effects on reproduction and development.

Therefore, be it resolved that Midway High School PTA supports legislation and actions that decrease, phase-out and eliminate the creation, release and exposure of dioxins.

Be it further resolved that MHS PTA supports the use of alternative processes, technologies, and products that avoid exposure to Dioxin, especially those that are chlorine-free.

Be it further resolved that this Dioxin Resolution is of state and national significance and shall be forwarded to Texas PTA and National PTA for consideration.

Contributors

L ois Marie Gibbs began her seventeen years of environmental activism as president of the Love Canal Homeowners Association in 1978. As founder and executive director, Lois has led CCHW's efforts to provide organizing, training, research, and educational and technical assistance services to communities in environmental crisis across the country for the past fourteen years. Her many honors include the 1990 Goldman Environmental Prize, *Outside* magazine's "Top Ten Who Made a Difference" Honor Roll in 1991, and an honorary Ph.D from the State University of New York, Cortland College.

Contributing Writers

C harlotte Brody is CCHW's organizing director. A registered nurse, she has worked on social justice issues since the 1960s. Before coming to CCHW, she was president and CEO of Planned Parenthood in Charlotte, North Carolina, and a founding staff member of the Brown Lung Association.

L in Kaatz Chary, MPH, is an environmental activist and doctoral student in public health at the University of Illinois at Chicago.

R ichard W. Clapp, Sc.D., M.P.H., is director of the Center for Environmental Health Studies JSI Research Training Institute in Boston, Massachusetts. His major research interests include environmental and cancer epidemiology and disease surveillance.

Charlie Cray is a toxics campaigner with Greenpeace in Chicago, Illinois.

Elizabeth Crowe is communications coordinator and community organizer with the Kentucky Environmental Foundation in Berea, Kentucky.

Alicia Culver is the coordinator of Ralph Nader's Government Purchasing Project (GPP) in Washington, D.C., which works to stimulate the markets for environmentally sound products and technologies using the purchasing power of government.

John Gayusky is research and organizing associate with the Citizens Clearinghouse for Hazardous Waste, and edits CCHW's quarterly magazine *Everyone's Backyard*.

Richard Grossman works with the Program on Corporations, Law and Democracy and is co-author of the pamphlet *Taking Care of Business: Citizenship and the Charter of Incorporation*.

Peter Kellman is a former president of Shoe Workers Local 82 and shop steward in the Painters Union. He worked for the Maine AFL-CIO and with the local unions in Jay during the International Paper strike of 1987-88. He lives in North Berwick, Maine.

Nina Laboy is the director of the Voter Participation Projects for the Community Service Society of New York. A longtime worker in the civil rights movement, Nina is a founding member of the National Congress for Puerto Rican Rights and the Bronx Clean Air Coalition. She lives in the South Bronx, New York.

Marilyn Leistner was the last mayor of Times Beach, Missouri, before the town was permanently evacuated. She has since led the fight to stop incineration of dioxin-contaminated soil from the former town site.

Stephen Lester, M.S., is science director with the Citizens Clearinghouse for Hazardous Waste. He has an M.S. in Toxicology from Harvard University and an M.S. in Environmental Health from New York University. Prior to joining CCHW, he served as technical advisor to the Love Canal Homeowner's Association.

Teresa Mills is the spokesperson for Parkridge Area Residents Take Action (PARTA) in Grove City, Ohio, and is the grassroots coordinator for the Ohio Environmental Council in Columbus.

Peter Montague, Ph.D., edits *Rachel's Environment & Health Weekly* (formerly *Rachel's Hazardous Waste News*), from Annapolis, Maryland.

David M. Ozonoff, M.D., MPH, is professor of Public Health and chair of the Department of Environmental Health at the Boston University School of Medicine. He is a licensed physician in the Commonwealth of Massachusetts.

Beverly J. Paigen, Ph.D., is senior staff scientist with the Jackson Laboratory in Bar Harbor, Maine.

Marvin Schneiderman, Ph.D., is a retired statistician and epidemiologist who spent over thirty years with the National Cancer Institute, where he was associate director for field studies.

Frank Solomon, Ph.D., is professor in the Department of Biology and Center for Cancer Research at the Massachusetts Institute of Technology in Cambridge, Massachusetts.

Terri Swearingen is a registered nurse, a mother, and a leading activist in the fight against the WTI incinerator in East Liverpool, Ohio. Terri was recently chosen by *Time Magazine* as one of the nation's top fifty promising leaders.

Joe Thornton is a research coordinator for the Greenpeace Toxics Campaign.

Carol Van Strum is a dioxin/PCB historian and co-founder of the citizens' group that won a landmark federal court ban of Agent Orange herbicides on national forests in 1977. The author of many publications on dioxin issues, she lives in Tidewater, Oregon.

Tom Webster is a doctoral student in the Department of Environmental Health at the Boston University School of Public Health. Tom spent eight years as a staff researcher for the Center for the Biology of Natural Systems and has authored many papers on dioxin.

Craig Williams is executive director of the Kentucky Environmental Foundation and national spokesperson for the Chemical Weapons Working Group, in Berea, Kentucky.

References

Abelson, P.H. (1994) "Chlorine and organochlorine compounds." *Science* 265: 1155, August 26.

(AE, 1994) *Permissible Political Activities, A Basic Guide for Grassroots Environmental Groups.* Americans for the Environment, July. Available from AE, 1400 16th Street, NW, Washington, D.C. 20036, 202/797-6665.

(AHA, 1995) Comments from the American Hospital Association on the USEPA dioxin exposure and health effects documents, January 13.

(AHA, 1995a) Letter written by Harry Bryan, Vice President, American Hospital Association, Division of Personal Membership Groups, to Judy Theisen, USEPA, January 13, as part of comments submitted on USEPA exposure and health effects documents.

Aittola, J., Paasivirta, J., and Vattulainen, A. (1992) "Measurements of organochloro compounds at a metal reclamation plant." Presented at Dioxin '92, twelfth International Symposium on Chlorinated Dioxins and Related Compounds, Tampere, Finland, August.

(AL, 1990) *Agent Orange: Vietnam Veterans Battle New Foes; Scientific Sleight-of-Hand, Bureaucratic Buck-Passing, Callous Coverup.* A report by the American Legion.

Allen, J.R., Barsotti, D.A., Lambrecht, L.K., and Van Miller, J.P. (1979) "Reproductive effects of halogenated aromatic hydrocarbons on nonhuman primates." *Annals of the New York Academy of Science* 320:419-25.

Allred, P.M. and Strange, J.R. (1977) "The effects of 2,4,5-trichlorophenoxyacetic acid and 2,3,7,8-tetrachlorodibenzo-p-dioxin on developing chicken embryos." *Archives of Environmental Contamination and Toxicology* 6:483-89.

Anthony, R.G., Garrett, M., and Schuler, C. (1993) "Environmental contaminants in bald eagles in the Columbia River Estuary." *Journal of Wildlife Management* 57:10-19.

(APHA, 1994) American Public Health Association Resolution 9304: "Recognizing and addressing the environmental and occupational health problems posed by chlorinated organic chemicals." *American Journal of Public Health* 84(3): 514-15.

Auger, J., Kunstmann, J.M., Gzyglik, F., and Jouannet, P. (1995) "Decline in semen quality among fertile men in Paris during the past 20 years." *New England Journal of Medicine* 332:281-85.

Bailey, J. (1992) "Dueling studies: how two industries created a fresh spin on the dioxin debate." *Wall Street Journal,* February 20.

Barsotti, D.A., Abrahamson, L.J., and Allen, J.R. (1979) "Hormonal alterations in female rhesus monkeys fed a diet containing 2,3,7,8-tetrachlorodibenzo-p-dioxin." *Bulletin of Environmental Contamination and Toxicology* 21:463-69.

Bates, M.N., Hannah, D.J., Buckland, S.J., Taucher, J.A., and van Mannen, T. (1994) "Chlorinated organic contaminants in breast milk of New Zealand women." *Environmental Health Perspectives* 102, Suppl. 1:211-17.

Beck, H., Dross, A., and Mathar, W. (1994) "PCDD and PCDF exposure and levels in humans in Germany." *Environmental Health Perspectives* 102, Suppl. 1:173-85.

Belkin, N.L. (1993) "Medical waste: reducing its generation." *Today's OR Nurse:* 40-42, September.

Berger, G.S. (1995) "Endometriosis: how many women are affected?" *Endometriosis Association Newsletter,* Vol. 14, No. 4: 4-7.

Bergeron, J.M., Crews, D., and McLachlan, J.A. (1994) "PCBs as environmental estrogens: turtle sex determination as a biomarker of environmental contamination." *Environmental Health Perspectives* 102:780-81.

Berry, R.M., Lutke, C.E., and Voss, R.H. (1993) "Ubiquitous nature of dioxins: a comparison of the dioxins content of common everyday materials with that of pulps and papers." *Environmental Science and Technology* 27(6):1164-68.

Bertazzi, P.A., Pesatori, A.C., Consonni, D., *et al.* (1993) "Cancer incidence in a population accidentally exposed to 2,3,7,8-tetrachlorodibenzo-para-dioxin." *Epidemiology* 4(5): 398-406

Bertazzi, P.A., Zocchetti C., Pesatori, A.C., *et al.* (1989) "Ten-year mortality study of the population involved in the Seveso incident in 1976." *American Journal of Epidemiology* 129:1187-1200.

Birnbaum, L. (1994) "The mechanism of dioxin toxicity: relationship to risk assessment." *Environmental Health Perspectives* 102, Suppl. 9:157-67, November.

Birnbaum, L. (1994a) "Endocrine effects of prenatal exposure to PCBs, dioxins, and other xenobiotics: implications for policy and future research." *Environmental Health Perspectives* 102, No. 8: 676-79, August.

Bobo, K., Kendall, J., and Max, S. (1991) *Organizing for Social Change: A Manual for Activists in the 1990s.* Cabin John, MD: Seven Locks Press.

Bond, G.G., McLaren, E.A., Brenner, F.E., and Cook, R.A. (1987) "Evaluation of mortality patterns among chemical workers with chloracne." *Chemosphere* 16:2117-21.

Booth, W. (1987) "Agent Orange study hits brick wall." *Science* 237:1285-86, September 11.

Bowman, R.E., Schantz, S.L., Weerasinghe, N.C.A., Gross, M.L., and Barsotti, D.A. (1989) "Chronic dietary intake of 2,3,7,8-tetrachlorodibenzo-p-dioxin (TCDD) at 5 or 25 parts per trillion in the monkey: TCDD kinetics and dose-effect estimate of reproductive toxicity." *Chemosphere* 18(1-6):243-52.

Bueno de Mesquita, H.B., Doornbos, G., van der Kuip, D.A.M., Kogevinas, M., and Winkelmann, R. (1993) "Occupational exposure to phenoxy herbicides and chlorophenols and cancer mortality in The Netherlands." *American Journal of Industrial Medicine* 23:289-300.

Calvert, G.M., Sweeney, M.H., Fingerhut, M.A., Hornung, R.W., and Halperin, W.E. (1992) "Hepatic and gastrointestinal effects in an occupational cohort exposed to 2,3,7,8-tetrachlorodibenzo-p-dioxin." *Journal of the American Medical Association* 267:2209-14.

Carey, J. and Reagan, M.B. (1995) "Are regs bleeding the economy?" *Business Week*:75-76, July 17.

Carlsen, E., Giwercman, A., Keiding, N., and Skakkebaek, N.E. (1992) "Evidence for decreasing quality of semen during past 50 years." *British Medical Journal* 305:609-13.

Casten, L.C. (1995) "EPA collusion with industry: a very brief overview." *The Orange Resource Book*, Don Fitz (editor), *Synthesis/Regeneration* 7/8. Published by WD Press, St. Louis, MO, summer. Available from Gateway Green Alliance, P.O. Box 8094, St. Louis, MO 63156, 314/727-8554.

Centers for Disease Control Veterans Health Studies (1988) "Serum 2,3,7,8-tetrachlorodibenzo-p-dioxin levels in U.S. Army Vietnam-era veterans." *Journal of the American Medical Association* 260:1249-54.

Centers for Disease Control Vietnam Experience Study (1988a) "Health status of Vietnam veterans." I. Psychosocial characteristics. *Journal of the American Medical Association* 259:2701-07.

Centers for Disease Control Vietnam Experience Study (1988b) "Health status of Vietnam veterans." II. Physical health. *Journal of the American Medical Association* 259:2708-14.

Centers for Disease Control Vietnam Experience Study (1988c) "Health status of Vietnam veterans." III. Reproductive outcomes and child health. *Journal of the American Medical Association* 259:2715-19.

Centers for Disease Control Vietnam Experience Study (1987) "Postservice mortality among Vietnam veterans." *Journal of the American Medical Association* 257:790-95.

(CFTA, 1995) Memorandum with attachments from Archie Beaton, Executive Director, Chlorine Free Trade Association, regarding list of TCF chemical pulp producers presented at International Non-Chlorine Bleaching Conference in March, plus additional listings received from U.S. and Canadian plants since the conference, August 29.

The Chemical Worker (1988) "Monsanto loses landmark dioxin case; jury awards $16.2 million to plaintiffs." January.

Chen, Y.C.J., Guo, Y.L.L., and Hsu, C.C. (1992) "Cognitive development of children prenatally exposed to polychlorinated biphenyls (Yu-Cheng children) and their siblings." *Journal of the Formosa Medical Association* 91: 704-7.

Cheung, M.O., Gilbert, E.F., and Peterson, R.E. (1981) "Cardiovascular teratogenicity of 2,3,7,8-tetrachlorodibenzo-p-dioxin in the chick embryo." *Journal of Toxicology and Applied Pharmacology* 61:197-204.

Cheung, M.O., Gilbert, E.F., and Peterson, R.E. (1981a) "Cardiovascular teratogenesis in chick embryos treated with 2,3,7,8-tetrachlorodibenzo-p-dioxin." In *Toxicology of Halogenated Hydrocarbons: Health and Ecological Effects*, Kahn, M.A.Q. and Stanton, R.H., eds. Elmsford, NY: Pergamon Press, 202-8.

Christian, R.S. (1994) Deputy Director for Research and Technology Assessment, American Legion, and Vice Chairman, Agent Orange Coordinating Council, "Comments to the U.S. Environmental Protection Agency public meeting on dioxin reassessment," December 8. Available from National Veterans Affairs and Rehabilitation Commission, American Legion, 1608 K Street NW, Washington, D.C. 20006.

Christian, R.S. (1992) Presentation at First Citizens Conference on Dioxin, Chapel Hill, NC, September 1991. Proceedings December 1992, 88-90. Available from Work On Waste USA, 82 Judson Street, Canton, NY 13617, 315/379-9200.

Christmann, W. (1989) "Combustion of polyvinyl chloride—an important source for the formation of PCDD/PCDF." *Chemosphere* 19:387-92.

Chriswell, C.D. (1977) "How much is a part per trillion, anyway?" *Chemecology* (letter), November.

Cohen, M., Commoner, B., *et al.* (1995) *Quantitative Estimation of the Entry of Dioxins, Furans and Hexachlorobenzene into the Great Lakes from Airborne and Waterborne Sources.* Center for the Biology of Natural Systems, Queens College, CUNY, Flushing, NY, May.

Cohen, M., (1995a) Personal communication, May.

Colborn, T., Dumanoski, D., and Meyers, J.P. (1996) *Our Stolen Future.* Dalton Press (in press).

Colborn, T., vom Saal, F.S., and Soto, A.M. (1993) "Developmental effects of endocrine-disrupting chemicals in wildlife and humans." *Environmental Health Perspectives* 101:378-84.

Colborn, T., *et al.* (1992) "Statement of consensus of the Wingspread work session: chemically-induced alterations in sexual and functional development: the wildlife/human connection." *Chemically-Induced Alterations in Sexual and Functional Development: The Wildlife/Human Connection.* Princeton, NJ: Princeton Scientific Publishers, 1-8.

Colborn, T., Davidson A., Green S.N., Hodge, R.A., Jackson, C.I., and Liroff, R.A. (1990) *Great Lakes, great legacy?* The Conservation Foundation, Washington, D.C.

Commoner, B. (1994) Presentation at the Second Citizens' Conference on Dioxin, "A turning point in the political history of dioxin," St. Louis, MO, July 30. Reprinted in *Waste Not,* August. Available from Work On Waste USA, 82 Judson Street, Canton, NY 13617, 315/379-9200.

Commoner, B. (1994a) "Comments on the EPA health assessment draft document for 2,3,7,8-tetrachlorodibenzo-p-dioxin (TCDD) and related compounds," December 12. Available from Center for the Biology of Natural Systems, Queens College, CUNY, Flushing, NY 11367, 718/670-4180 or fax, 718/670-4189.

Constable, J.D. and Hatch, M.C. (1985) "Reproductive effects of herbicide exposure in Vietnam: recent studies by the Vietnamese and others." *Teratogenesis, Carcinogenesis, and Mutagenesis* 5:231-50.

Costner, P., *et al.* (1995) *PVC: A Primary Contributor to the U.S. Dioxin Burden.* Greenpeace International, February.

Costner, P. (1995a) *The Incineration of Dioxin-Contaminated Wastes in Jacksonville, Arkansas: A Review of the Inhalation Exposure Assessment.* Greenpeace International, August 7.

Courtney, K.D. and Moore, J.A. (1971) "Teratology studies with 2,4,5-T and 2,3,7,8-TCDD." *Journal of Toxicology and Applied Pharmacology* 20:396-403.

Culotta, E. (1995) "New evidence about feminized alligators." *Science* 267:330-31.

Czuczwa, J.M. and Hites, R.A. (1986) "Airborne dioxins and dibenzofurans: sources and fates." *Environmental Science and Technology* 20:195-200.

Czuczwa, J.M. and Hites, R.A. (1985) "Historical record of polychlorinated dioxins and furans in Lake Huron sediments." In *Chlorinated Dioxins and Dibenzofurans in the Total Environment II*, Keith, L.H., Rappe, C., and Choudhary, G., eds. Boston, MA: Butterworth Publishers.

Czuczwa, J.M., McVeety, B.D., and Hites, R.A. (1984) "Polychlorinated dibenzo-p-dioxins and dibenzofurans in sediments from Siskiwit Lake, Isle Royale." *Science* 226: 568-69.

Davis, D.L., Bradlow, H.L., Wolff, M., Woodruff, T., Hoel, D.G., and Anton-Culver, H. (1993) "Medical hypothesis: xenoestrogens as preventable causes of breast cancer." *Environmental Health Perspectives* 101:372-77.

Della Porta, G., Dragani, T.A., and Sozzi, G. (1987) "Carcinogenic effects of infantile and long-term 2,3,7,8-tetrachlorodibenzo-p-dioxin treatment in the mouse." *Tumori* 73: 99-107.

Demassa, D.A., Smith, E.R., Tennent, B., and Davidson, J.M. (1977) "The relationship between circulating testosterone levels and male sexual behavior in rats." *Hormones and Behavior* 8: 275-86.

Dempsey, C.R. and Oppelt, E.T. (1993) "Incineration of hazardous waste: a critical review update." *Air & Waste* 43:25-73.

Devesa, S.S., Blot, W.J., Stone, B.J., *et al.* (1995) "Recent cancer trends in the United States." *Journal of the National Cancer Institute* 87:175-82.

DeVito, M.J. and Birnbaum, L.S. (1994) "Toxicology of dioxins and related chemicals." In *Dioxins and Health*, Arnold Schecter, ed., NY: Plenum Press, 139-62.

Dewailly, E., Ryan, J.J., Laliberte, C., *et al.* (1994) "Exposure of remote maritime populations to coplanar PCBs." *Environmental Health Perspectives* 102 Suppl. 1:205-9.

DiGiacomo, J.C., Odom, J.W., Ritota, P.C., and Swam, K.G. (1992) "Cost containment in the operating room: use of reusable versus disposable clothing." *American Surgeon* 5(1):653-57.

(DOH, 1981) *Love Canal, A Special Report to the Governor and Legislature*. New York State Department of Health.

Donna, A., Crosignani, P., Robutti, F., et al. (1989) "Triazine herbicides and ovarian epithelial neoplasms." *Scandinavian Journal of Work, Environment and Health* 15:47-53.

Donna, A., Betta, P.G., Robutti, F., Crosignani, P., Berrino, F., and Bellingeri, D. (1984) "Ovarian mesothelial tumors and herbicides: a case-control study." *Carcinogenesis* 5:941-42.

(Dow, 1978) *The Trace Chemistries of Fire—A Source of and Routes for the Entry of Chlorinated Dioxins into the Environment.* Published by the Chlorinated Dioxin Task Force, Michigan Division, Dow Chemical U.S.A.

Egeland, G.M., Sweeney, M.H., Fingerhut, M.A., Wille, K.K., Schnorr, T.M., and Halperin, W.E. (1994) "Total serum testosterone and gonadotropin in workers exposed to dioxin." *American Journal of Epidemiology* 139: 272-81.

Environmental Research (1988) Editorial and five epidemiological studies. Vol. 47, No. 2, December.

Erickson, J.D., Mulinare, J., McClain, P.W., Fitch, T.G., James, L.M., McClearn, A.B., and Adams, M.J., Jr. (1984) "Vietnam veterans risks for fathering babies with birth defects." *Journal of the American Medical Association* 252:903-12.

Eriksson, M., Hardell, L., and Adami, H.O. (1990) "Exposure to dioxins as a risk factor for soft tissue sarcoma: a population-based case-control study." *Journal of the National Cancer Institute* 82: 486-90.

Evans, G.R., Webb, K.B., Knutsen, A.P., Roodman, S.T., Roberts, D.W., and Garrett, W.A. (1988) "A medical follow-up of the health effects of long-term exposure to 2,3,7,8-tetrachlorodibenzo-p-dioxin." *Archives of Environmental Health* 43: 273-78.

Falck, F., Jr., Ricci, A., Jr., and Wolff, M.S., Godbold, J., and Deckers, P. (1992) "Pesticides and polychlorinated biphenyl residues in human breast lipids and their relation to breast cancer." *Archives of Environmental Health* 47:143-46.

Farland, W. (1995) "Status of dioxin-related activities at the United States Environmental Protection Agency (EPA)." *Health & Environment Digest*, Vol. 9, No. 2, May/June.

Fernandez-Salguero, P., Pineau, T., Hilbert, D.M., et al. (1995) "Immune system impairment and hepatic fibrosis in mice lacking the dioxin-binding Ah receptor." *Science* 268:722-26.

Fett, M.J., Adena, M.A., Cobbin, D.M., and Dunn, M. (1987) "Mortality among Australian conscripts of the Vietnam conflict era." I. Causes of death. *American Journal of Epidemiology* 125:878-84.

Fingerhut, M.A., Halperin, W.E., Marlow, D.A., *et al.* (1991) "Cancer mortality in workers exposed to 2,3,7,8-tetrachlorodibenzo-p-dioxin." *New England Journal of Medicine* 324:212-18.

Fitrakis, B. (1994) "Smoking guns and sinking stacks: how SWACO rigged the trash test." *Free Press* (Columbus, OH), summer, 3-4.

Fox, J.L. (1984) "Agent Orange study is like a chameleon." *Science* 223: 1156-57, March 16.

Freeman, R.A., Hileman, F.D., Noble, R.W., and Schroy, J.M. (1987) "Experiments on the mobility of 2,3,7,8-tetrachlorodibenzo-p-dioxin at Times Beach." In *Solving Hazardous Waste Problems*, Exner, J.H., ed. ACS symposium series, No. 338.

French, H.M. (1994) "Blueprint for reducing, reusing, and recycling." *Association of Operating Room Nurses Journal* 60(1):94-98.

Furst, P., Furst, C., and Wilmers, K. (1994) "Human milk as a bioindicator for body burden of PCDDs, PCDFs, organochlorine pesticides, and PCBs." *Environmental Health Perspectives* 102, Suppl. 1:187-93.

(GPP, 1995) *Dioxin-Free Products Guide.* Ralph Nader's Government Purchasing Project (GPP) provides a detailed list of products that create dioxin in their manufacture, use or disposal as well as safer alternatives available to government purchasing officials and other consumers. This book also presents model ordinances (from the U.S.A. and abroad) and numerous studies of government agencies that have successfully utilized dioxin-free substitutes. Available from GPP ($10), P.O. Box 19367, Washington, D.C. 20036.

Giwercman, A., Carlsen, E., Keiding, N., and Skakkebaek, N.E. (1993) "Evidence for increasing incidence of abnormalities of the human testis: a review." *Environmental Health Perspectives* 101, Suppl. 2:65-71.

Goldman, B. (1994) *Toxic Waste and Race Revisited.* Center for Policy Alternatives, Washington, D.C.

Goldman, P.J. (1972) "Critically acute chloracne caused by trichlorophenol decomposition products." *Arbeitsmed. Sozialmed. Arbeitshygiene* 7:12-18.

Gorman, C. (1991) "The double take on dioxin." *Time* 52: August 26.

Gray, L.E., Jr., Kelce, W., Monosson, E., Ostby, J.S., and Birnbaum, L.S. (1995) "Perinatal TCDD reduces ejaculated and epididymal sperm counts in rats and hamsters, but does not affect testosterone or androgen receptor levels." *Journal of Toxicology and Applied Pharmacology* (in press).

Gray, L.E., Ostby, J.S., Kelce, W., Marshall, R., Diliberto, J.J., Birnbaum, L.S. (1993) "Perinatal TCDD exposure alters sex differentiation in both female and male LE hooded rats." Abstracts from Dioxin '93, thirteenth International Symposium on Chlorinated Dioxins and Related Compounds, Vienna, 337-39.

Greenpeace (1995) "Pulp and Paper Fact Sheet." Available from Greenpeace U.S.A., 1436 U Street NW, Washington, D.C. 20009, 202/462-1177.

Greenpeace (1994) *Fraudulent and/or Invalid Human Studies on Effects of TCDD*. Petition before the Administrator, USEPA 28 U.S.C. 2620.

Greenpeace International (1993) *Dioxin Factories: A Study of the Creation and Discharge of Dioxins and Other Organochlorines from the Production of PVC*. Amsterdam, The Netherlands: Greenpeace International.

Greenpeace International (1995) "Chlorine crisis: pulp fact, pulp and paper 2."

Greig, J.B., Jones, G., Butler, W.H., and Barnes, J.M. (1973) "Toxic effects of 2,3,7,8-tetrachlorodibenzo-p-dioxin." *Food and Cosmetics Toxicology* 11:585-95.

Griffith, J., Duncan, R., Riggan, W.B., and Pellom, A.C. (1989) "Cancer mortality in U.S. counties with hazardous waste sites and groundwater pollution." *Archives of Environmental Health* 44:69-74.

Grossman, R.L. and Frank, T.A. (1993) *Taking Care of Business: Citizenship and the Charter of Incorporation*. Cambridge, MA: Charter, Inc.

Gruber, L., Stohrer, E., and Santi, H. (1993) "PCDD/F in the paper industry: new results and mass balances." Thirteenth International Symposium on Chlorinated Dioxins and Related Compounds. *Organohalogens* 11:315-18.

Guillette, L.J., Jr., Gross, T.S., Masson, G.R., Matter, J.M., Percival, H.F., and Woodward, A.R. (1994) "Developmental abnormalities of the gonad and abnormal sex hormone concentrations in juvenile alligators from contaminated and control lakes in Florida." *Environmental Health Perspectives* 102:680-88.

Guo, Y.L., Lai, T.J., Ju, S.H., Chen, Y.C., and Hsu, C.C. (1993) "Sexual developments and biological findings in Yu-Cheng children." Presented at Dioxin '93, thirteenth International Symposium on Chlorinated Dioxins and Related Compounds, Vienna, Austria, September 20-24.

Guo, Y.L., Lin, C.J., Yao, W.J., and Hsu, C.C. (1992) "Musculoskeletal changes in Yu-Cheng children compared with their matched controls." Proceedings of the twelfth International Symposium on Dioxins and Related Compounds, Tampere, Finland.

Gupta, P.K. and Gupta, R.C. (1979) "Pharmacology, toxicology and degradation of endosulfan." *Toxicology* 13:115-30.

Guzelian, P.S. (1982) "Comparative toxicology of chlordecone (kepone) in humans and experimental animals." *Annual Review of Pharmacology and Toxicology* 22:89-113.

Hall, B. (1988) *Environmental Politics: Lessons from the Grassroots.* Durham, NC: Institute for Southern Studies.

Hatch, M. (1984) *Herbicides and war: the long-term ecological and human consequences,* Westing, A.H., ed. Philadelphia, PA: Taylor and Francis.

Hay, A. (1992) Presentation at First Citizens Conference on Dioxin, Chapel Hill, NC, September 1991. Proceedings December 1992, 101-3.

(HED, 1992) "Dioxins & their cousins: the Dioxin '91 conference." *Health & Environment Digest,* Vol. 5, No. 12, March.

Heindl, A. and Hutzinger, O. (1987) "Search for industrial sources of PCDD/PCDF—III. Short-chain chlorinated hydrocarbons." *Chemosphere* 16:1949-57.

Hiatt, F. (1983) "Air Force reports 'first good news' on Agent Orange." *The Washington Post,* A10, July 2.

Hickman, D., Chang, D., and Glasser, H. (1989) "Cadmium and lead in biomedical waste incinerators." Presented at eighty-second annual meeting of the air and waste management association, Anaheim, CA, June.

Hilts, P. (1987) "Veterans and Agent Orange: a debate that eludes scientific resolution." *The Washington Post,* A11, September 28.

Hoffman, R.E., Stehr-Green, P.A., Webb, K.B., *et al.* (1986) "Health effects of long-term exposure to 2,3,7,8-tetrachlorodibenzo-p-dioxin." *Journal of the American Medical Association* 255:2031-38.

Holden, C. (1979) "Agent Orange furor continues to build." *Science* 205:770-72, August 24.

Homberger, E., Reggiani, G., Sambeth, J., and Wipf, H.K. (1979) "The Seveso accident: its nature, extent and consequences." *Annals of Occupational Hygiene,* Vol. 22:327-70.

Horstmann, M., McLachlan, M.S. (1994) "Textiles as a source of polychlorinated dibenzo-p-dioxins and dibenzofurans (PCDD/CDF) in human skin and sewage sludge." *Environmental Science & Pollution Resource* 1(1):15-20.

Houk, V.H. (1989) Testimony of Vernon H. Houk, M.D., Director, Center for Environmental Health and Injury Control, Centers for Disease Control, Department of Health and Human Services, before the U.S. House of Representatives, Committee on Government Operations, Human Resources and Intergovernmental Relations Subcommittee, July 11.

Hutzinger, O. and Fiedler, H. (1991) "Formation of dioxins and related compounds in industrial processes." In *Dioxin Perspectives: A Pilot Study on International Information Exchange on Dioxins and Related Compounds,* Bretthauer, E.W., Kraus, H.W., and di Domenico, A. eds. New York, NY: Plenum Press.

(HWN, 1992) "New evidence that all landfills leak." *Rachel's Hazardous Waste News* #316, December 16. Available from Environmental Research Foundation, P.O. Box 5036, Annapolis, MD 22403-7036, 410/263-1584.

(HWN, 1991) "Plastics—part 2: why plastic landfill liners always fail." *Rachel's Hazardous Waste News* #217, January 23. Available from Environmental Research Foundation, P.O. Box 5036, Annapolis, MD 22403-7036, 410/263-1584.

(HWN, 1990) "Fine particles—part 5: incineration worsens landfill hazards." *Rachel's Hazardous Waste News* #162, January 3. Available from Environmental Research Foundation, P.O. Box 5036, Annapolis, MD 22403-7036, 410/263-1584.

(IJC, 1994) *Seventh Biennial Report on Great Lakes Water Quality,* International Joint Commission, ed. Washington, D.C. and Ottawa, Ontario.

Inside EPA Weekly Report (1987) "ORD effort expected to yield lower EPA risk assessment of dioxin." Vol. 8, No. 43, October 23.

Irvine, D. (1994) "Falling sperm quality." *British Medical Journal* 309:476, August 13.

Jenkins, C. (1991) *Recent Scientific Evidence Developed After 1984 Supporting a Causal Relationship Between Dioxin and Human Health Effects.* United States District Court for the Eastern District of New York. CV-89-03361 (E.D.N.Y.) (JBW) [B-89-00559-CA (E.D. TEX.)]

Jenkins, C. (1990) "Newly revealed fraud by Monsanto in an epidemiological study used by EPA to assess human health effects from dioxins." Memorandum to Ray C. Loehr, Chair, Executive Committee, Science Advisory Board (A 101), USEPA, Office of the Administrator, from Cate Jenkins, Ph.D., Chemist, Waste Characterization Branch (OS 332), Characterization and Assessment Branch, February 23.

Kahn, P.C., Gochfield, M., Nygren, M., Hansson, M., Rappe, C., Velez, H., Ghent-Guenther, R.N., and Wilson, W.P. (1988) "Dioxins and dibenzofurans in blood and adipose tissue of Agent Orange-exposed Vietnam veterans and matched controls." *Journal of the American Medical Association* 259:1661-67.

Kemner, et al. v. Monsanto (1989) Appeal, Fifth Appellable Court, Illinois, No. 5:88-420.

Khera, K.S. and Ruddick, J.A. (1973) "Polychlorinated dibenzo-p-dioxins: perinatal effects and the dominant lethal test in Wistar rats." *Advances in Chemistry Series 120*, E.H. Blair, ed. American Chemical Society 120:70-84.

Kleeman, J.M., Olson., J.R., and Peterson, R.E. (1988) "Species differences in 2,3,7,8-tetrachlorodibenzo-p-dioxin toxicity and biotransformation in fish." *Fundamentals of Applied Toxicology* 10:206-13.

Kociba, R.J., Keyes, D.G., Beyer, J.E., Carreon, R.M., Wade, C.E., Dittenber, D.A., Kalnins, R.P., Frauson, L.E., and Park, C.N. (1978) "Results of a two-year chronic toxicity and oncogenicity study of 2,3,7,8-tetrachlorodibenzo-p-dioxin in rats." *Journal of Toxicology and Applied Pharmacology* 46:279-303.

Kociba, R.J., Keeler, P.A., Park, G.N., and Gehring, P.J. (1976) "2,3,7,8-tetrachlorodibenzo-p-dioxin (TCDD): results of a 13-week oral toxicity study in rats." *Journal of Toxicology and Applied Pharmacology* 35:553-74.

Kuratsune, M. (1989) "Yusho, with reference to Yu-Cheng." In *Halogenated biphenyls, terphenyls, napthalenes, dibenzodioxins and related products*, Kimbrough, R.D. and Jensen, AA., eds. New York, NY: Elsevier Science Publishers, 2nd ed., 381-400.

Kuratsune, M., Ikeda, M., Nakamura, Y. and Hirohata, T. (1988) "A cohort study on mortality of Yusho patients: a preliminary report." In *Unusual Occurrences as Clues to Cancer Etiology*. Miller, R.W., *et al.*, eds. Tokyo, Japan: Japan Sci. Soc. Press/Taylor & Francis, Ltd., 61-68.

Kuratsune, M., Ikeda, M., Nakamura, Y., Hirohata, T. (1987) "A cohort study on mortality of "Yusho" patients: a preliminary report." *International Symposium Princess Takamatsu Cancer Research Fund* 18:61-66.

Laber (1994) "Burning of trash infuriates neighbors." *The Record* (Hackensack, NJ), November 18.

LaFleur, L., Bousquet, T., Ranage, K., *et al.* (1990) "Analysis of TCDD and TCDF on the ppq-level in milk and food sources." *Chemosphere* 20 (10-12):1657-62.

Lahl, U. (1994) "Sintering plants of steel industry-PCDD/F emission status and perspectives." *Chemosphere*, Vol. 29, Nos. 9-11, 1939-49.

Lerda, D. and Rizzi, R. (1991) "Study of reproductive function in persons occupationally exposed to 2,4-dichlorophenoxyacetic acid (2,4-D)." *Mutation Research* 262:47-50.

Ligon, W.V., Jr., Dorn, S. B., and May, R. J. (1989) "Chlorodibenzofuran and Chlorodibenzo-p-dioxin levels in chilean mummies dated to about 2800 years before the present." *Environmental Science and Technology* 23, No. 10:1286-90.

Mably, T.A., Moore, R.W., and Peterson, R.E. (1992) "In utero and lactational exposure of male rats to 2,3,7,8-tetrachlorodibenzo-p-dioxin: 1. effects on androgenic status." *Journal of Toxicology and Applied Pharmacology* 114:97-107.

Mably, T.A., Moore, R.W., Goy, R.W., and Peterson, R.E. (1992a) "In utero and lactational exposure of male rats to 2,3,7,8-tetrachorodibenzo-p-dioxin: 2. Effects on sexual behavior and the regulation of luteinizing hormone secretion in adulthood." *Journal of Toxicology and Applied Pharmacology* 114:108-17.

Mably, T.A., Bjerke, D.L., Moore, R.W., Gendron-Fitzpatrick, A., and Peterson, R.E. (1992b) "In utero and lactational exposure of male rats to 2,3,7,8-tetrachlorodibenzo-p-dioxin: 3. effects on spermatogenesis and reproductive capability." *Journal of Toxicology and Applied Pharmacology* 114: 118-26.

Mably, T.A., Moore, R.W., Bjerke, D.L., and Peterson, R.E. (1991) "The male reproductive system is highly sensitive to in utero and lactational 2,3,7,8-tetrachlorodibenzo-p-dioxin exposure." In *Biological Basis for Risk Assessment of Dioxins and Related Compounds.* Gallo, M.A, Sceuplein, R.J., and van der Heijden, C.A., eds., Banbury Report 35. Cold Spring Harbor, NY: Cold Spring Harbor Laboratory, 69-78.

Madhukar, B.V., Brewster, D.W., and Matsumura, F. (1984) "Effects of in vivo-administered 2,3,7,8-tetrachlorodibenzo-p-dioxin on receptor binding of epidermal growth factor in the hepatic plasma membrane of rat, guinea pig, mouse, and hamster." *Proceedings of the National Academy of Sciences* (USA) 81: 7404-11.

Manz, A., Berger, J., Dwyer, J.H., Flesch-Janys, D., Nagel, S., and Waltsgott, H. (1991) "Cancer mortality among workers in a chemical plant contaminated with dioxin." *Lancet* 338:959-64.

Marshall, E. (1993) "Search for a killer: focus shifts from fat to hormones." *Science* 259:618-21, January 29.

Martin, J.V. (1984) "Lipid abnormalities in workers exposed to dioxin." *British Journal of Medicine* 41:254-256.

(MDEP, 1994) *State of Maine 1994 Water Quality Assessment*. Maine Department of Environmental Protection, Augusta, ME.

Melius, J.M., Lewis-Michl, E.L., Kallenback, L.R., *et al*. (1994) *Residence near industries and high traffic areas and the risk of breast cancer on Long Island*. New York State Department of Health, April.

Merrell, P.E. (1992) Presentation at First Citizens Conference on Dioxin, Chapel Hill, NC, September 1991. Proceedings December 1992, 108-9. Available from Work On Waste USA, 82 Judson Street, Canton, NY 13617, 315/379-9200.

Michalek, J.E., Wolfe, W.H., and Miner, J.C., (1990) "Health status of Air Force veterans occupationally exposed to herbicides in Vietnam. II. mortality." *Journal of the American Medical Association* 264:1832-36.

Mocarelli, P., Marocchi, A., Brambilla, P., Gerthoux, P.M., Young, D.S., and Mantel, N. (1986) "Clinical laboratory manifestations of exposure to dioxin in children. A six-year study of the effects of an environmental disaster near Seveso, Italy." *Journal of the American Medical Association* 256:2687-95.

Monks, V. (1993) "See No Evil." *American Journalism Review*, 18-25, June.

Moses, M., Lilis, R., Crow, K.D., Thornton, J., Fischbein, A., Anderson, H.A., and Selikoff, I.J. (1984) "Health status of workers with past exposure to 2,3,7,8-tetrachlorodibenzo-p-dioxin in the manufacture of 2,4,5-trichlorophenoxyacetic acid. Comparison of findings with and without chloracne." *American Journal of Industrial Medicine* 5:161-82.

Mueller, T. (1994) "Back to the reusable future." *Biocycle*, 36. February.

Murray, F.J., Smith, F.A., Nitschke, K.D., Humiston, C.G., Kociba, R.J., and Schwetz, B.A. (1979) "Three-generation reproduction study of rats given 2,3,7,8-tetrachlorodibenzo-p-dioxin (TCDD) in the diet." *Journal of Toxicology and Applied Pharmacology* 50:241-52.

Mussalo-Rauhamaa, H., Hasanen, E., Pyysalo, H., Antervo, K., Kauppila, R., and Pantzar, P. (1990) "Occurrence of betahexachlorocyclohexane in breast cancer patients." *Cancer* 66:2124-28.

Muto, H. and Takizawa, Y. (1989) "Dioxins in cigarette smoke." *Archives of Environmental Health* 44 (3):171-74, May/June.

Nadler, R.D. (1969) "Differentiation of the capacity for male sexual behavior in the rat." *Hormones and Behavior* 1:53-63.

National Academy of Sciences (1993) *Health Effects of the Herbicides Used in Vietnam*, Institute of Medicine. Washington, D.C.: National Academy Press.

National Toxicology Program (NTP) (1982) *Carcinogenesis Bioassay of 2,3,7,8-tetrachlorodibenzo-p-dioxin in Osborne-Mendel Rats and B6C3F1 Mice (Gavage Study)*. Technical report series, No. 209. DHEW Publication No. (NIH) 82-1765.

(NCASI, 1993) *Summary of Data Reflective of Pulp and Paper Industry Progress in Reducing the TCDD/TCDF Content of Effluent, Pulps and Wastewater Treatment Sludges*. National Council of the Paper Industry for Air and Stream Improvement, New York, NY, June.

(NELC, 1993) *An Ounce of Toxic Pollution Prevention: State Toxics Use Reduction Laws*. National Environmental Law Center, and Center for Policy Alternative, 2nd ed., January.

Neubert, D. and Dillman, I. (1972) "Embryotoxic effects in mice treated with 2,4,5-trichlorophenoxyacetic acid and 2,3,7,8-tetrachlorodibenzo-p-dioxin." *Archives of Pharmacology* 272:243-64.

Newman, P. (1994) *Communities at Risk: Contaminated Communities Speak Out on Superfund*. The Center for Community Action and Environmental Justice, Riverside, CA.

Oehme, M., Mano, S., and Bjerke, B. (1989) "Formation of polychlorinated dibenzofurans and dibenzo-p-dioxins by production processes for magnesium and refined nickel." *Chemosphere* 18(7-8):1379-89.

Ohio Environmental Protection Agency (OEPA, 1994) "Risk assessment of potential health effects of dioxins and dibenzofurans emitted from the Columbus Solid Waste Authority's reduction facility." February 28.

Orband, J.E., Stanley, J.S., Schwemberger, J.G., and Remmers, J.C. (1994) "Dioxins and dibenzofurans in adipose tissue of the general U.S. population and selected subpopulations." *American Journal of Public Health* 84:439-45.

Ozonoff, D. (1995) Letter to Morton Lippmann, Chair, USEPA Science Advisory Board Dioxin Reassessment Review Committee, from David Ozonoff, Boston University School of Public Health in the School of Medicine, Department of Environmental Health, May 26.

Papke, O., Ball, M., Lis, Z.A., and Scheunert, K. (1989) "PCDD/PCDF in whole blood samples of unexpected persons." *Chemosphere* 19:941-48.

Patrick Engineering (1994) *Western Lake Superior Sanitary District, Burn Barrel Dioxin Test*, August 1992.

Patterson, D.G., Jr., Todd, G.D., Turner, W.E., Maggio, V., Alexander, L.R., and Needham, L.L., (1994) "Levels of non-ortho-substituted (coplanar), mono- and di-ortho-substituted polychlorinated biphenyls, dibenzo-p-dioxins, and dibenzofurans in human serum and adipose tissue." *Environmental Health Perspectives* 102, Suppl. 1:195-204.

Pazderova-Vejlupkova, J., Nemcova, M., Pickova, J., Jirasek, L., and Lukas, E. (1981) "The development and prognosis of chronic intoxication by tetrachlorodibenzo-p-dioxin in man." *Archives of Environmental Health* 36:5-11.

Peterson, R.E., Theobald, H.M., and Kimmel, G.L. (1993) "Developmental and reproductive toxicity of dioxins and related compounds: cross-species comparisons." *Critical Reviews of Toxicology* 23:283-335.

(PSR, 1995) Letter from Alan H. Lockwood, M.D., Physicians for Social Responsibility, to USEPA.

Pinter, A., Torok, G., Borzsonye, M., Surjan, A., Calk, M., Kelecsenvi, Z., and Kocsis, Z. (1990) "Long-term carcinogenicity bioassay of the herbicide atrazine in F344 rats." *Neoplasma* 37:533-44.

Pirkle, J.L., Wolfe, W.H., and Patterson, D.G., *et al.* (1989) "Estimates of the half-life of 2,3,7,8-TCDD in Vietnam veterans of Operation Ranch Hand." *Journal of Toxicology and Environmental Health* 27:165-71.

Pluim, H.J., Boersma, R., Kramer, L., Vander, D., Slikke, J.W., and Koppe, J.G. (1994) "Influence of short-term dietary measures on dioxin concentration in human milk." *Environmental Health Perspectives* 102:968-71.

Pluim, H.J., Koope, J.G., Olie, K., *et al.* (1992) "Effects of dioxins on the thyroid function in newborn babies." *Lancet* 339:1303.

Porterfield, S.P. (1994) "Vulnerability of the developing brain to thyroid abnormalities; environmental insults to the thyroid system." *Environmental Health Perspectives* 102, Suppl. 2:125-30.

(PTCN, 1985) "Pesticides possibly contaminated with halogenated dibenzo-p-dioxins by USEPA Office of Pesticide Programs." *Pesticide and Toxic Chemical News* 13(15):34-38, February 20.

Rabe, A. and Nelson, L. (1992) "Building a movement for environmental justice." *Everyone's Backyard*, Vol. 10, No. 5, October.

Raloff, J., (1994) "What's in a cigarette?" *Science News* 145:330-31, May 21.

Rao, M.S., Subbarao, V., Prasad, J.D., and Scarpelli, D.C. (1988) "Carcinogenicity of 2,3,7,8-tetrachlorodibenzo-p-dioxin in the Syrian golden hamster." *Carcinogenesis* 9(9): 1677-79.

Rawls, R.L. (1979) "Dow finds support, doubt for dioxin ideas." *Chemical & Engineering News*: 23-29, February 12.

Reggiani, G. (1980) "Acute human exposure to TCDD in Seveso, Italy." *Journal of Toxicology and Environmental Health* 6:27-43.

Reggiani, G. (1978) "Medical problems raised by the TCDD contamination in Seveso, Italy." *Archives of Toxicology* 40: 161-88.

Rier, S.E., Martin, D.C., Bowman, R.E., Dmowski, W.P., and Becker, J.L. (1993) "Endometriosis in rhesus monkeys (*macaca mulatta*) following chronic exposure to 2,3,7,8-tetrachlorodibenzo-p-dioxin." *Fundamentals of Applied Toxicology* 21:433-41.

Riggle, D. (1994) "Advanced hospital recycling." *Biocycle*, February.

Roberts, L. (1991) "Flap erupts over dioxin meeting." *Science* 251:866-67, February 22.

Roberts, L. (1991a) "EPA moves to reassess the risk of dioxin." *Science* 252:911, May 17.

Roberts, L. (1991b) "More pieces in the dioxin puzzle." *Science* 254:377, October 18.

Roegner, R.H., Grubbs, W.D., Lustik, M.B., *et al.* (1991) *Air Force Health Study: An Epidemiological Investigation of Health Effects in Air Force Personnel Following Exposure to Herbicides.* Serum dioxin analysis of 1987 examination results. NTIS# AD A-237-516 through AD A-237-524.

Rogan, W.J., Gladen, B.C., Hung, K.L., *et al.* (1988) "Cogenital poisoning by polychlorinated biphenyls and their contaminants in Taiwan." *Science* 241:334-38.

Rogan, W.J., Gladen, B.C., McKinney, J., *et al.* (1987) "Polychlorinated biphenyls (PCBs) and dichlorodiphenyl dichloroetene (DDE) in human milk; effects on growth, morbidity and duration of lactation." *American Journal of Public Health* 77:1294-97.

Rogan, W.J., Gladen, B.C., McKinney, J., Carreras, N., Hardy, P., Thullen, J., Tingelstad, J., and Tully, M. (1986) "PCBs and DDE in human milk." *American Journal of Public Health* 76:172-77.

Rohleder, R. (1989) "Dioxins and cancer mortality—reanalysis of the BASF cohort." Paper presented at Dioxin '89 Ninth International Symposium on Chlorinated Dioxins and Related Compounds, Toronto, Ontario, Canada, September 12-22. *Chemosphere* (in press).

Rosenberg, H. (1995) "Summary of visit to Seveso, Italy." April 27.

Ross, P., De Swart, R., Reijinders, P., Van Loveren, H., Vos, J., and Osterhaus, A. (1995) "Contaminant-related suppression of delayed-type hypersensitivity and antibody responses in harbor seals fed herring from the Baltic Sea." *Environmental Health Perspectives* 103:162-67.

Rossberg, M., *et al.* (1986) "Chlorinated hydrocarbons." *Encyclopedia of Industrial Chemistry*, Gerhartz, W., ed., 5th ed. New York, NY: VCH Publishers, 233-398.

Sanjour, W. (1994) "The Monsanto Investigation." Memorandum to David Bussard, Director, Characterization and Assessment Division, from Policy Analyst, Office of Solid Waste, July 20.

Sanjour, W. (1994a) "Columbus, Ohio incinerator." Follow-up memorandum to Carol Browner, Administrator, USEPA, from Policy Analyst, Office of Solid Waste, February 22.

Sanjour, W. (1994b) "Columbus, Ohio incinerator." Memorandum to Valdus V. Adamkus, Regional Administrator, Region V, from Policy Analyst, Office of Solid Waste, April 12.

Santii, R., Newbold, R.R., Makela, S., Pylkkanen, L., and McLachlan, J.A. (1994) "Development estrogenization and prostatic neoplasia." *Prostate* 24:67-78.

Saracci, R., Kogevinas, M., Bertazzi, P., *et al.* (1991) "Cancer mortality in an international cohort of workers exposed to chlorophenoxy herbicides and chlorophenols." *Lancet* 338: 1027-32.

Schantz, S.L., Laughlin, M.K., Van Valkenberg, H.C., Bowman, R.E. (1986) "Maternal care by rhesus monkeys of infant monkeys exposed to either lead or 2,3,7,8-tetrachlorodibenzo-p-dioxin (TCDD)." *Neurotoxicology* 7:641-54.

Scharnberg, K. (1992) "Searching for a new start: America's homeless veterans, firebase dioxin." *The American Legion*, Vol. 132, No. 2, 30-31, 59, February.

Schecter, A, ed. (1994) *Dioxins and Health*, Binghamton, NY: Plenum Press.

Schecter, A., Startin, J., Wright, C., Kelly, M., Papke, O., Lis, A., Ball, M., and Olson, J.R. (1994a) "Congener-specific levels of dioxins and dibenzo-furans in U.S. food and estimated daily dioxin toxic equivalent intake." *Environmental Health Perspectives* 102:962-66.

Schecter, A., Furst, P., Furst, C., and Papke, O., *et al.* (1994b) "Chlorinated dioxins and dibenzofurans in human tissue from general populations: a selective review." *Environmental Health Perspectives* 102, Suppl. 1:159-71.

Schecter, A., Ryan, JJ., Masuda, Y., *et al.* (1994c) "Chlorinated and brominated dioxins and dibenzofurans in human tissue following exposure." *Environmental Health Perspectives* 102, Suppl. 1:135-47.

Schecter, A., Stanley, J., Boggess, K., Masuda, Y., Mes, J., Wolff, M., Furst, P., Furst, C., Wilson-Yang, K., and Chisholm, B. (1994d) "Polychlorinated biphenyl levels in the tissue of the exposed and nonexposed humans." *Environmental Health Perspectives* 102, Suppl. 1:149-58.

Schecter, A., di Domenico, A., Tirrio-Baldassarri, L., and Ryan, J. (1992) "Dioxin and dibenzofuran levels in milk of women from four geographical regions in Italy as compared to levels in other countries." Presented at: Dioxin '92, Twelfth International Symposium on Chlorinated Dioxins and Related Compounds, Tampere, Finland, August.

Schecter, A., Papke, O., and Ball, M. (1990) "Chlorinated dioxin and dibenzofuran levels in human blood samples from various countries, including Vietnam, the Soviet Union, the United States and Germany." In *Organohalogen Compounds* (Dioxin '90), Vol. 1, 271-73. Bayreuth, Germany: Ecoinforma Press.

Schecter, A., Dekin, A., Weerasinghe, N.C.A., Arghestani, S., and Gross, M.L. (1987) "Source of dioxins in the environment: study of PCDDs and PCDFs in ancient, frozen Eskimo tissue." *Chemosphere* 17:627-31.

Schneider, K. (1991) "U.S. officials say dangers of dioxin were exaggerated." *New York Times*, front page, August 15.

Science News. (1990) "Biomedicine: Vietnam veterans sustain cancer threat." Vol. 137:236.

Scribner, J.D. and Mottet, N.K. (1981) "DDT acceleration of mammary gland tumors induced in the male Sprague-Dawley rat by 2-acetamidophenanthrene." *Carcinogenesis* 2:1235-39.

Sharpe, R.M. and Skakkebaek, N.E. (1993) "Are estrogens involved in falling sperm count and disorders of the male reproductive tract?" *Lancet* 341:1392-95.

Sim, M.R., and McNeil, J.J. (1992) "Monitoring chemical exposure using breast milk; a methodological review." *American Journal of Epidemiology* 136:1-11.

Singh, S.K. and Pandey, R.S. (1990) "Effect of sub-chronic endosulfan exposures on plasma gonadotrophins, testosterone, testicular testosterone, and enzymes of androgen biosynthesis in rat." *International Journal of Experimental Biology* 28:953-56.

Smith, D.W. (1995) "Are PCBs in the Great Lakes approaching a 'new Equilibrium'?" *Environmental Science & Technology* 29 (1):42A-46A.

Smith, A.H., Fisher, D.O., *et al.* (1984) "Soft tissue sarcoma and exposure to phenoxyherbicides and chlorophenols in New Zealand." *Journal of the National Cancer Institute* 73:1111-17.

Smith, A.H., Fisher, D.O., Giles, H.J., *et al.* (1983) "The New Zealand soft tissue sarcoma case-control study: interview findings concerning phenoxyacetic acid exposure." *Chemosphere* 12:565-71.

Smith, J. (1979) "EPA halts most use of herbicide 2,4,5-T." *Science* 203:1090-91, March 6.

Sodersten, P. and Hansen, S. (1978) "Effects of castration and testosterone, dihydrostestosterone or oestradiol replacement treatment in neonatal rats on mounting behavior in the adult." *Journal of Endocrinology* 76: 251-60.

Stellman, S.D., Stellman, J.M., and Sommer, J.F. (1988) "Health and reproductive outcomes among American Legionnaires in relation to combat and herbicide exposure in Vietnam." *Environmental Research* 47 (2): 150-74, December.

Stockbauer, J.W., Hoffman, R.E., Schramm, W.F., and Edmonds, L.D. (1988) "Reproductive outcomes of mothers with potential exposure to 2,3,7,8-tetrachlorodibenzo-p-dioxin." *American Journal of Epidemiology* 128:410-19.

Stone, R. (1995) "Dioxin receptor knocked out." *Science* 268:638-39, May 5.

Stone, R. (1995a) "PCBs pack hormonal punch." *Science* 267:1770-71, March 24.

Suskind, R.R. and Hertzberg, V.S. (1984) "Human health effects of 2,4,5-T and its toxic contaminants." *Journal of the American Medical Association* 251:2372-80.

Swedish Environmental Protection Agency (1994) *Sample Reports #610s09 and 0610s017: PVC Suspension/PVC Plastics and PVC Suspension used to make PVC.* May 5.

Sweeney, M.H., Hornung, R.W., Wall, D.K., Fingerhut, M.A., and Halperin, W.E. (1992) "Prevalence of diabetes and increased fasting serum glucose in workers with long-term exposure to 2,3,7,8-tetrachlorodibenzo-p-dioxin." Presented at Twelfth International Symposium on Dioxins and Related Compounds, Tampere, Finland, August 24-28.

Sweeney, M.H., Fingerhut, M.A., Connally, L.B., Halperin, W.E., Moody, P.L., and Marlowe, D.A. (1989) "Progress of the NIOSH cross-sectional medical study of workers occupationally exposed to chemicals contaminated with 2,3,7,8-TCDD." *Chemosphere* 19:973-77.

Thiessen, J., *et al.* (1989) "Determination of PCDFs and PCDDs in fire accidents and laboratory combustion tests involving PVC-containing materials." *Chemosphere* 19:423-28.

Thornton, J. (1995) "Comments on USEPA dioxin reassessment: estimating exposures to dioxin-like compounds," Greenpeace toxics campaign. January 13.

Thornton, J. (1994) *Achieving Zero Dioxin: An Emergency Strategy for Dioxin Elimination.* Available from Greenpeace Chlorine-Free Campaign, 1436 U Street NW, Washington, D.C. 20009. July.

Tieszen, M.E. and Gruenberg, J.C. (1992) "A quantitative, qualitative and critical assessment of surgical waste." *Journal of the American Medical Association* 267(20):2765-68.

Tilson, H.A., Davis, G.J., McLachlan, J.A., and Lucier, G.W. (1979) "The effects of polychlorinated biphenyls given prenatally on the neurobehavioral development of mice." *Environmental Research* 18:466-74.

(TMC, 1994) *The People, The Press & Politics: The New Political Landscape.* News release, Times Mirror Center for the People & the Press, Washington, D.C.

(USAF, 1988) *United States Air Force Personnel and Exposure to Herbicide Orange.* USAF School of Aerospace Medicine, Human Systems Division (AFSC), Brooks Air Force Base, TX. USAFSAM-TR-88-3 interim report for period March 1984-February 1988.

U.S. Department of Commerce (1993) *Statistical Abstract of the United States 1992.* Bureau of the Census, Washington, D.C.

(USEPA, 1995) *Proposed Standards and Guidelines for Medical Incinerators.* 60 Federal Register 10654. February 27.

(USEPA, 1994) *Health Assessment Document for 2,3,7,8-tetrachlorodibenzo-p-dioxin (TCDD) and Related Compounds.* Vol. I of III, USEPA, Office of Research and Development, EPA/600/BP-92/001a, external review draft, June.

(USEPA, 1994a) *Health Assessment Document for 2,3,7,8-tetrachlorodibenzo-p-dioxin (TCDD) and Related Compounds,* Vol. II of III, USEPA, Office of Research and Development, EPA/600/BP-92/001b, external review draft, June.

(USEPA, 1994b) *Health Assessment Document for 2,3,7,8-tetrachlorodibenzo-p-dioxin (TCDD) and Related Compounds,* Vol. III of III, USEPA, Office of Research and Development, EPA/600/BP-92/001c, external review draft, June.

(USEPA, 1994c) *Estimating Exposure to Dioxin-Like Compounds.* Vol. I: Executive Summary, USEPA, Office of Research and Development, EPA/600/6-88/005Ca, external review draft, June.

345

(USEPA, 1994d) *Estimating Exposure to Dioxin-Like Compounds*, Vol. II: Properties, Sources, Occurrence and Background Exposures, USEPA, Office of Research and Development, EPA/600/6-88/005Cb, external review draft, June.

(USEPA, 1994e) *Estimating Exposure to Dioxin-Like Compounds*, Volume III: Site-Specific Assessment Procedures, USEPA, Office of Research and Development, EPA/600/6-88/005Cc, external review draft, June.

(USEPA, 1994f) "EPA calls for new dioxin data to complete reassessment process." USEPA, Environmental News Communications, Education, and Public Affairs (1703), September 13.

(USEPA, 1993) *Locating and Estimating Air Emissions from Sources of Dioxins and Furans.* Draft final report. Prepared by Radian Corporation for the Office of Air Quality Planning and Standards, Research Triangle Park, NC. September 30.

(USEPA, 1993a) *Development Document for Proposed Effluent Limitations Guidelines and Standards for the Pulp, Paper and Paperboard Point Source Category.* Office of Water, EPA/821/R/93/019.

(USEPA, 1987) *National Dioxin Study, Report to Congress.* Office of Solid Waste and Emergency Response, EPA/530-SW-87-025, August.

(USEPA, 1985) *Health Assessment Document for Polychorinated Dibenzo-p-dioxins.* EPA Office of Health and Environmental Assessment, EPA/600-8-84/014f, September.

(USEPA, 1983) *Dioxin Strategy.* Prepared by the Office of Water Regulations and Standards and the Office of Solid Waste and Emergency Response in conjunction with the Dioxin Strategy Task Force, Washington, D.C., November 28.

(USEPA, 1982) *Federal Register Notice: National Toxicology Program Announces Availablity of Cancer Biossay Reports on 2,3,7,8-tetrachlorodibenzo-p-dioxin (Dermal and Gavage Studies) and Locust Bean Gum.* FR:47(No. 37):8094, February 24.

(USEPA, 1979) *Report of Assessment of a Field Investigation of Six-Year Spontaneous Abortion Rates in Three Oregon Areas in Relation to Forest 2,4,5-T Spray Practice.* February.

U.S. General Accounting Office (GAO 1994) *Hazardous Waste: Issues Pertaining to an Incinerator in East Liverpool, Ohio. Resources, Community, and Economic Development Division.* GAO/RCED-94-101, September 9.

U.S. General Accounting Office (GAO 1986) *Agent Orange: VA Needs to Further Improve Its Examination and Registry Program.* GAO/HRD-86-7, January 14.

U.S. International Trade Commission (1991) *Synthetic Organic Chemicals—United States Production, 1990.* USITC Public No. 2470, Washington, D.C.

U.S. Office of Technology Assessment (1990). *Finding the Rx for Managing Medical Wastes.* OTA-0-459, U.S. Government Printing Office, Washington, D.C.

Van Miller, J.P., Lalich, J.J., and Allen, J.R. (1977) "Increased incidence of neoplasms in rats exposed to low levels of 2,3,7,8-tetrachlorodibenzo-p-dioxin." *Chemosphere* 10:625-32.

Van Strum, C. (1992) Presentation at First Citizens Conference on Dioxin, Chapel Hill, NC, September 1991. Proceedings December 1992. 104-6.

Van Strum, C. and Merrell, P. (1987) *No Margin of Safety: A Preliminary Report on Dioxin Pollution and the Need for Emergency Action in the Pulp and Paper Industry,* Greenpeace USA, Inc.

Walker, M.K. and Peterson, R.E. (1991) "Potencies of polychlorinated dibenzo-p-dioxins, dibenzofurans, and biphenyl congeners for producing early life stage mortality in rainbow trout (Oncorhyncus mykiss)." *Aquatic Toxicology* 21: 219-38.

Waste Not (1991) "That benign little carcinogen called dioxin." Work on Waste USA, 82 Judson Street, Canton, NY 13617, 315/379-9200, August 29. Also available as a 10-volume videotape series of the First Citizens' Conference on Dioxin, September 21-22. Available from Video-Active Productions, Rt. 2, Box 322, Canton, NY 13617, 315/386-8797.

Webb, K.B., Evans., R.G., Knudsen, D.P., and Roodman, S. (1989) "Medical evaluation of subjects with known body levels of 2,3,7,8-tetrachlorodibenzo-p-dioxin." *Journal of Toxicology and Environmental Health* 28:183-93.

Webster, T. and Commoner, B. (1994) "The dioxin debate: overview." In *Dioxins and Health,* Schecter, A., ed., New York, NY: Plenum Press, 1-50.

Weiss, T. (1989) Oversight review of CDC's Agent Orange study, opening statement before the U.S. House of Representatives, Human Resources and Intergovernmental Relations Subcommittee of the Committee on Government Operations, Washington, D.C. July 11.

Weisskopf, M. (1987) "Paper industry campaign defused reaction to dioxin contamination." *Washington Post,* October 25.

Weisskopf, M. (1987a) "Dioxin found in some paper products: concentrations reported in EPA study seen as no threat to health." *Washington Post,* September 25.

Westin, J.B. and Richter, E. (1990) "The Israeli breast-cancer anomaly in trends in cancer mortality in industrial countries." *Annals of the New York Academy of Sciences* 609:269-79.

Whorton, M.D. and Foliart, D.E. (1983) "Mutagenicity, carcinogenicity and reproductive effects of dichloropropane (DBCP)." *Mutation Research* 123:13-30.

Wiemeyer, S., Lamont, T., Bunck, S., Sindelar, C., Gramlich, F., Fraser, J., and Byrd, M. (1984) "Organochlorine pesticide, polychlorobiphenyl, and mercury residues in bald eagles, 1969-1979, and their relationships to shell thinning and reproduction." *Archives of Environmental Contamination and Toxicology* 13:529-49.

Wolff, M.S., Paolo, G., Toniolo, P., Lee, E.W., Rivera, M., and Dubin, N. (1993) "Blood levels of organochlorine residues and risk of breast cancer." *Journal of National Cancer Institute* 85:648-52.

Wolfe, W.H., Michalek, J.E., Miner, J.C., and Rahe, A.J. (1992) *Air Force Health Study: An Epidemiologic Investigation of Health Effects in Air Force Personnel Following Exposure to Herbicides. Reproductive Outcomes.* Brooks Air Force Base, TX, Epidemiology Research Division, Armstrong Laboratory, Human Systems Division (AFSC).

Wolfe, W.H., Michalek, J.E., Miner, J.C., *et al.* (1990) "Health status of Air Force veterans occupationally exposed to herbicides in Vietnam." I. Physical Health. *Journal of the American Medical Association* 264(14):1824-31.

Yanchinski, S. (1989) "New analysis links dioxin to cancer." *New Scientist*, 24. October 28.

Yost, P. (1989) "Agent Orange study called botched or rigged," *Washington Post*, July 12.

Zack, J.A. and Suskind, R.R. (1980) "The mortality experience of workers exposed to tetrachlorodibenzo-p-dioxin in a trichlorophenol process accident." *Journal of Occupational Medicine* 22(1):11-14.

Zober, A., Messerer, P., and Huber, P. (1990) "Thirty-four-year mortality follow-up of BASF employees exposed to 2,3,7,8-TCDD after the 1953 accident." *International Archives of Occupational and Environmental Health* 62:139-57.

Zumwalt, E.R. (1995) "Binding up the wounds." In *Dioxin: The Orange Resource Book, Synthesis/Regeneration 7/8*, summer. Available from Gateway Green Alliance, P.O. Box 8094, St. Louis, MO 63156 ($7).

Index

Brominated substances; Furans; Polychlorinated biphenyls

Dioxin Resolution, sample of, 319-20

Dioxins, 297: environmental fate of, 41-44; physical properties of, 40-41; toxicity of, 301

Direct action, 182-84; civil disobedience, 184-85; letter-writing, 183; nonviolent, 182; plan, 181-83, 184; to stop all incineration, 239-42; to stop chlorine use and production, 247-50; to stop medical waste incinceration, 245-47. *See also* Activism; Organizing

Dow Chemical Company, xxix, 3, 8-10; and cancer study, 25; and class action lawsuit, 18-19; and "dioxin from fire" theory, xxx, 9-11, 56; new products, 289; *Trace Chemistries of Fire*, 9; *See also* 2,4,5-T; Agent Orange

Downwinders, 261

Drums, reclamation of, 235

Durbin, Irene (and family), 122-24

Dyes, 224-25

E

East Liverpool (Ohio), 86. *See also* Waste Technologies Incorporated

ECF. *See* Elemental chlorine free

Electoral politics, 273-76

Electrical equipment, 234-35

Elemental chlorine free (ECF), 249, 266

El Pueblo Para El Aire Y Aqua Limpia (People for Clean Air and Water), 251

Emission estimates, 48-49, 58-62

Endometriosis. *See* Reproduction

Environmental Health Monthly (Citizens Clearinghouse for Hazardous Waste), xxiv

Environmental justice, 250-53. *See also* Principles of Environmental Justice

Environmental Politics: Lessons from the Grassroots (Hall), 193-94

Environmental Protection Agency (EPA): and buried evidence, 122-25; vs. community' groups, 149-50; as participant in poisoning, 148; as permitter of poisoning, 146-48; "public review draft" ("EPA report," 1994), xviii, xxxi, 23-24, 30-32, 47; reassessment, first (1988), xxxi, 27-29; reassessment, second (1991), 29-30; regulatory efforts, 3; risk assessment (1985), xxxi, 25-27; workgroup, 27-28. *See also* Greenpeace; Office of Criminal Investigations; Pulp and paper industry; Science Advisory Board

Environmental racism. *See* Racism

EPA. *See* Environmental Protection Agency

"EPA report." *See* Environmental Protection Agency

Escambia Treating Company (Pensacola, Florida), 57, 252-53

"Estrogens in the Environment," 105

Ethylene dichloride, 220-21

Everyone's Backyard (Citizens Clearinghouse for Hazardous Waste), xxiv

I

J

K

L

M

N

O

MANUFACTURING CONSENT

Noam Chomsky And the Media
Mark Achbar, ed.

2nd printing

Manufacturing Consent Noam Chomsky and the Media, the companion book to the award-winning film, charts the life of America's most famous dissident, from his boyhood days running his uncle's newsstand in Manhattan to his current role as outspoken social critic.

A complete transcript of the film is complemented by key excerpts from the writings, interviews and correspondence. Also included are exchanges between Chomsky and his critics, historical and biographical material, filmmakers' notes, a resource guide and more than 270 stills from the film.

A juicily subversive biographical/philosophical documentary bristling and buzzing with ideas.
Washington Post
You will see the whole sweep of the most challenging critic in modern political thought.
Boston Globe
One of our real geniuses... an excellent introduction.
Village Voice
An intellectually challenging crash course in the man's cooly contentious analysis, laying out his thoughts in a package that is clever and accessible.
Los Angeles Times
... challenging, controversial... the unravelling of ideas.
Globe and Mail
...lucid and coherent statement of Chomsky's thesis.
The Times of London
... invaluable as a record of a thinker's progress towards basic truth and basic decency.
The Guardian

264 pages, 270 illustrations, bibliography, index
Paperback ISBN: 1-551640-02-3
 $19.99 within Canada
$23.99 outside Canada
Hardcover ISBN: 1-551640-03-1
 $48.99 within Canada
$52.99 outside Canada

COMMUNITY ACTION

Organizing for Social Change

Henri Lamoureux, Robert Mayer and Jean Panet-Raymond

trans. by Phyllis Aronoff and Howard Scott

... a thoroughly readable and immensely useful work... required reading for community activists.
Quill & Quire

248 pages, bibliography
Paperback ISBN: 0-921689-20-9 $16.99
Hardcover ISBN: 0-921689-21-7 $45.99

FIGHTING FOR HOPE

Organizing to Realize Our Dreams

Joan Newman-Kuyek

This book finds the common threads in activism and weaves them together into a guidebook for social change.

Kuyek provides a valuable list of do's and don'ts for social-justice activists, and her emphasis on building structures for self-reliant communities is important for today's victims of recession.
Calgary Herald

221 pages
Paperback ISBN: 0-921689-86-1 $17.99
Hardcover ISBN: 0-921689-87-X $46.99
L.C. No. 90-83629

MYTH OF THE MARKET

Promises and Illusions

Jeremy Seabrook

The majority of the people place their hope and faith in the mechanism of the market as the bearer of promise for the future, but the spreading of market values leads to social disintegration and the destruction of cultures.

A strong indictment of the market system. All the more timely with the recent moves in global trade.
Peace and Environment News
There are alternatives to the market, but unless we begin to resist the monetization of all human activity, they will be relegated to museums where, fittingly, you'll pay to see them.
Imprint

189 pages
Paperback ISBN: 1-895431-08-5 $18.99
Hardcover ISBN: 1-895431-09-3 $47.99

ECOLOGY OF FREEDOM

The Emergence and Dissolution of Hierarchy, *revised edition*
Murray Bookchin

The most systematic articulation of ideas.
San Francisco Review of Books
... a confirmation of his [Bookchin's] status as a penetrating critic not only of the ways in which humankind is destroying itself, but of the ethical imperative to live better.
Village Voice
Elegantly written, and recommended for a wide audience.
Library Journal

395 pages, index
Paperback ISBN: 0-921689-72-1 $19.99
Hardcover ISBN: 0-921689-73-X $48.99
L.C. No. 90-83628

COMMON CENTS

Media Portrayal of the Gulf War and Other Events
James Winter

Objectivity is the theme of these five case studies which deal with how the media covered the Gulf War, the Oka standoff, the Ontario NDP's budget, the Meech Lake Accord and Free Trade.

Winter provides strong evidence of a corporate tilt in the mass media... it is impossible to dismiss [his] arguments.
Vancouver Sun

Like Chomsky, he enjoys contrasting the "common-sense" interpretation with views from alternative sources. As facts and images clash, we end up with a better grasp of the issues at hand.
Montréal Gazette

304 pages, index
Paperback ISBN: 1-895431-24-7 $23.99
Hardcover ISBN: 1-895431-25-5 $52.99

WORDS OF A REBEL

Peter Kropotkin

First published during Kropotkin's imprisonment, this collection contains articles written between 1879 and 1882. The first complete English version.

229 pages
Paperback ISBN: 1-895431-04-2 $19.99
Hardcover ISBN: 1-895431-05-0 $48.99

PASSION FOR RADIO

Radio Waves and Community
Bruce Girard, ed.

A project of the World Association of Community Radio Broadcasters, this book tells the stories of alternative radio projects around the globe, from First Nations in the Canadian North, to punks in Amsterdam, progressives in California, and guerrillas in El Salvador.

The stories in this book are both moving and inspiring.
Media Development

This impressive book is an exciting window into the increasingly diffuse world of participatory media.
Media Information Australia

212 pages
Paperback ISBN: 1-895431-34-4 $19.99
Hardcover ISBN: 1-895431-35-2 $48.99
L.C. No. 91-72979

BANKERS, BAGMEN, AND BANDITS

Business and Politics in the Age of Greed
R.T. Naylor

Based on Naylor's widely read column, this book is designed to give the news behind the news, to put back into the stories the 'awkward' details the main stream media find convenient to omit.

An eminently readable book, with outré insights into the corrupt underside of world affairs in each chapter.
Canadian Book Review Annual

250 pages
Paperback ISBN: 0-921689-76-4 $18.99
Hardcover ISBN: 0-921689-77-2 $47.99
L.C. No. 90-83630

PRIVATE INTEREST, PUBLIC SPENDING
Balanced-Budget Conservatism and the Fiscal Crisis
William Scheuerman and Sidney Plotkin

Extremely informative about the political trends that exist and about the political economic system of business dealing with government.
The Activist

280 pages, index
Paperback ISBN: 1-895431-98-0 $19.99
Hardcover ISBN: 1-895431-99-9 $48.99

BUREAUCRACY AND COMMUNITY
Linda Davies, Eric Shragge, eds.

This book examines the consequences for both State social workers and community practitioners in face of increasing governmental restraints.

Based on recent empirical work from Québec and the United Kingdom.

... *takes a highly critical view of social-services management and the controlling role of government bureaucracies.*
Calgary Herald

180 pages, bibliography
Paperback ISBN: 0-921689-56-X $16.99
Hardcover ISBN: 0-921689-57-8 $45.99

YEAR 501
The Conquest Continues
Noam Chomsky

2nd printing
A powerful and comprehensive discussion of the incredible injustices hidden in our history.

... Year 501 *offers a savage critique of the new world order.*
MacLean's Magazine
Tough, didactic, [Chomsky] skins back the lies of those who make decisions.
Globe and Mail
... *a much-needed defense against the mind-numbing free market rhetoric.*
Latin America Connexions
331 pages, index
Paperback ISBN: 1-895431-62-X $19.99
Hardcover ISBN: 1-895431-63-8 $48.99

ELECTRIC RIVERS
The Story of the James Bay Project
Sean McCutcheon

... *a book about how and why the James Bay project is being built, how it works, the consequences its building will have for people and for the environment... it cuts through the rhetoric so frequently found in the debate.*
Canadian Book Review Annual
Electric Rivers *is a welcome contribution to the debate... a good fortune for readers who would like to better understand a story that is destined to dominate the environmental and political agenda.*
Globe and Mail

194 pages, maps
Paperback ISBN: 1-895431-18-2 $18.99
Hardcover ISBN: 1-895431-19-0 $47.99
L.C. No. 91-72981

INDIGNANT HEART

A Black Worker's Journal
Charles Denby

A two-part account of a U.S. activist's battle for freedom, first personally and then as a supporter of the principal movements of the last twenty-five years.

295 pages
Paperback ISBN: 0-919618-67-7 $9.99
Hardcover ISBN: 0-919618-93-6 $38.99

BETWEEN THE LINES

How to Detect Bias and Propaganda in the News and Everyday Life
Eleanor MacLean

Taking professional journalism to task for not practising fully enough the lofty ideals it preaches.
Canadian Journal of Communication
An excellent resource tool for teachers.
Kingston Whig-Standard

296 pages
Paperback ISBN: 0-919619-12-6 $19.99
Hardcover ISBN: 0-919619-14-2 $48.99

BEYOND HYPOCRISY

Decoding the News in an Age of Propaganda
Including a Doublespeak Dictionary for the 1990s

Edward S. Herman

Illustrations by Matt Wuerker

In a highly original volume that includes an extended essay on the Orwellian use of language that characterizes U.S. political culture, cartoons, and a cross-referenced lexicon of *doublespeak* terms with examples of their all too frequent usage, Herman and Wuerker highlight the deception and hypocrisy contained in the U.S. government's favourite buzz-words.

Rich in irony and relentlessly forthright, Beyond Hypocrisy *is a valuable resource for those interested in avoiding... 'an unending series of victories over your own memory'.*
Montréal Mirror
Edward Herman starts out with a good idea and offers a hard-hitting and often telling critique of American public life.
Ottawa Citizen

239 pages, illustrations, index
Paperback ISBN: 1-895431-48-4 $19.99
Hardcover ISBN: 1-895431-49-2 $48.99

**BLACK
ROSE
BOOKS**

has also published the following books of related interest

Politics of Sustainable Development: Citizens, Unions and the
   Corporations, *by Laurie E. Adkin*
Anarchism and Ecology, *by Graham Purchase*
Ecology as Politics, *by André Gorz*
Green Politics: Agenda for a Free Society, *by Dimitrios Roussopolous*
Ecology of the Automobile, *by Peter Freund and George Martin*
Europe's Green Alternative, *edited by Penny Kemp*
No Nukes: Everyone's Guide to Nuclear Power, *by Anna Gyorgy and Friends*
Nuclear Power Game, *by Ronald Babin*
Sun Betrayed: A Study of the Corporate Seizure of Solar Energy
   Development, *by Ray Reece*
Free Trade: Neither Free Nor About Trade, *by Christopher Merrett*
First Person Plural: A Community Development Approach to Social
   Change, *by David Smith*
Services and Circuses: Community and the Welfare State,
   by Frédéric Lesemann, translated by Lorne Huston and Margaret Heap
The Search for Community: From Utopia to a Cooperative Society,
   by George Melnyk
Who is this "We"? Absence of Community, *by Eleanor Godway and
   Geraldine Finn*
Local Places: In the Age of the Global City, *edited by Roger Keil,
   David V.J. Bell and Gerda R. Wekerle*
Bringing the Economy Home From the Market, *by Ross Dobson*

send for a free catalogue of all our titles
BLACK ROSE BOOKS
C.P. 1258
Succ. Place du Parc
Montréal, Québec
H3W 2R3 Canada

To order books in North America: (phone) 1-800-565-9523
(fax) 1-800-221-9985
In Europe: (phone) 44-081-986-4854 (fax) 44-081-533-5821

Web site address: http://www.web.net/blackrosebooks

Printed by the workers of
Les Éditions Marquis
Montmagny, Québec